T0191947

Engineering Materials

This series provides topical information on innovative, structural and functional materials and composites with applications in optical, electrical, mechanical, civil, aeronautical, medical, bio- and nano-engineering. The individual volumes are complete, comprehensive monographs covering the structure, properties, manufacturing process and applications of these materials. This multidisciplinary series is devoted to professionals, students and all those interested in the latest developments in the Materials Science field, that look for a carefully selected collection of high quality review articles on their respective field of expertise.

More information about this series at http://www.springer.com/series/4288

Liwei Wang · Guoliang An · Jiao Gao ·
Ruzhu Wang

Property and Energy Conversion Technology of Solid Composite Sorbents

Liwei Wang
Mechanical Engineering School
Shanghai Jiao Tong University
Shanghai, China

Guoliang An
Mechanical Engineering School
Shanghai Jiao Tong University
Shanghai, China

Jiao Gao
Mechanical Engineering School
Shanghai Jiao Tong University
Shanghai, China

Ruzhu Wang
Mechanical Engineering School
Shanghai Jiao Tong University
Shanghai, China

ISSN 1612-1317   ISSN 1868-1212 (electronic)
Engineering Materials
ISBN 978-981-33-6090-7   ISBN 978-981-33-6088-4   (eBook)
https://doi.org/10.1007/978-981-33-6088-4

Jointly published with Science Press
The print edition is not for sale in China (Mainland). Customers from China (Mainland) please order the print book from: Science Press.

This Springer imprint is published by the registered company Springer Nature Singapore Pte Ltd.
The registered company address is: 152 Beach Road, #21-01/04 Gateway East, Singapore 189721, Singapore

Preface

Sustainable development is the common pursuit for the people worldwide, for which efficient energy utilization and environmental protection are the key strategies.

Solid sorption technology is driven by low-grade thermal energy and has been used for energy storage, refrigeration, and heat pump. Compared with compression type refrigerator/heat pump, solid sorption technology has advantages for energy saving because the thermal energy input is generally provided by solar energy, waste heat from factories, and geothermal heat, etc. It is also a type of environmentally benign technology for utilizing refrigerants with zero ODP (ozone depletion potential) and GWP (Global Warming Potential). Compared with absorption technology also driven by the low-grade thermal energy, the solid sorption has wide scopes of candidate sorbents, including different physical and chemical sorbents. Such a merit could fit different application occasions with the low-grade heat in the range of 50–300 °C. Moreover, the solid sorption technology doesn't need the solution pump or rectification equipment, neither refrigerant pollution nor solution crystallization exists which are the intractable problems for the liquid sorption technology. So solid sorption technology has been recognized as one of the most useful approaches to renewable energy utilization. The solid sorbents widely used in solid sorption technology include activated carbons (ACs), silica gel, zeolites, oxides, and halides. Most of the solid sorption cycles are discontinuous batch processes involving sorption and desorption phases. To fulfill an efficient cycle, one of the major technical challenges is to achieve high heat transfer performance with reasonable mass transfer performance of the system.

The main technologies researched by the academics on solid sorption are advanced solid sorbents, cycles, and system design. Among them, the development of outstanding solid sorbents is the most significant part. In our previous monograph (Adsorption Refrigeration: Theory and Applications, Published by Wiley in 2014), researches on physical, chemical, and composite sorbents were summarized, and the advanced cycles, such as heat recovery cycle, mass recovery cycle, and thermal wave cycle were analyzed in detail. The applications of solid sorption technology for using solar energy and waste heat from engines were also introduced. But in that book, the introduction to composite sorbents was general and not summarized in detail. Furthermore, it had a little discussion on the composite sorption kinetics and comprehensive applications of the composite solid sorbents.

In recent years, we have gotten a few achievements on the development, kinetics study, and applications of composite solid sorbents. Our studies show that composite sorbents, especially the composite sorbents with expanded natural graphite (ENG) as the matrix, have excellent heat and mass transfer performance, good stability for sorption and desorption, and can make the size of the sorbent bed compact. Based on the composite solid sorbents, we have designed solid sorption systems for refrigerated vehicles, energy storage, and electricity generation. In order to show the detailed development and applications of composite solid sorbents to readers, in this book we summarized the researches on relating contents.

One feature of the composite solid sorbent is the strong anisotropic heat and mass transfer property. For example, the layered structure of CENG (compact expanded natural graphite) improves the gas diffusion in the direction perpendicular to that of compression, while the continuous structure of the ENG also makes this direction become the optimum direction for the heat transfer. The kinetic studies of the composite sorbents show a very interesting phenomenon, that is the sorption and desorption processes under non-equilibrium conditions are bi-variable other than mono-variable, which seems like a feature of physical adsorption. The multi-halide sorbents are analyzed in this book, and by proper proportioning, such type of sorbents could make the hysteresis phenomena attenuate, which is firstly found for solid sorption processes. Because the water refrigerant has the feature of low (partial) pressure in sorption and desorption processes, for water uptake consolidated solid sorbent, the mass transfer should be considered thoroughly. Due to this reason, a new type of structure with wave type layers is proposed and analyzed.

Consolidated composite sorbents have excellent performance for many applications. In this book, the composite sorbents have been used for the refrigerated truck, whose freezing system is driven by the waste gas from the engine and could save a significant part of the energy for keeping the frozen food inside the cabin. The composite sorbents have also been utilized for energy storage, and the studies show the resorption process has competitive performance. Similarly, the resorption electricity cogeneration process is firstly analyzed, which has better performance than that of sorption/desorption processes. The air-to-water system used for recovering solar energy to produce water is set up and has potential utilization for the desert and island where less water exists. A new type of de-NOx cycle and system for trucks is also proposed in this book.

All the applications of composite solid sorbents are hot research topics nowadays. This book could give readers a comprehensive guide on the development of composite sorbents and relating systems.

Shanghai, China

Liwei Wang
Guoliang An
Jiao Gao
Ruzhu Wang

Acknowledgements

We are grateful for the contributions of academics and students in our research team, they are: Dr. K. Daou, Dr. L. Jiang, Dr. J. Y. Wang, and B. Tian, who contributed to the development of solid composite sorbents; Dr. L. Jiang, Dr. X. Zheng, Z. Q. Jin, and K. Wang, who contributed to the properties of solid composite sorbents; Z. S. Zhou, who contributed to the kinetics of solid composite sorbents; Dr. P. Gao, Dr. J. Y. Wang, Dr. Y. Yu, J. Wang, and Z. X. Wang, who contributed to the solid sorption cycle for refrigeration, water production, heat transfer, and eliminating NOx emission; Dr. H. S. Bao, Dr. L. Jiang, and F. Q. Zhu, who contributed to the solid sorption cycle for energy storage, electricity generation, and cogeneration; Y. C. Tian, C. Zhang, Y. H. Zhang, and S. F. Wu, who contributed to the revision of this book.

We also appreciate the support from the Foundation for National Natural Science Foundation of China for the Distinguished Young Scholars (Grant No. 51825602), Innovative Research Groups of the National Natural Science Foundation of China (Grant No. 51521004), and the National Natural Science Foundation of China for Excellent Young Scholars (Grant No. 51222601).

Contents

About the Authors

Liwei Wang is a Professor of Mechanical Engineering School in Shanghai Jiao Tong University. She has long focused on the solid sorption technology and novel thermodynamic cycles driven by the low-grade thermal energy. She developed the new type solid composite single-halide and multi-halide sorbents, which solved the performance attenuation of chemisorption working pairs, decreased the chemisorption hysteresis successfully, and improved the thermal conductivity of granular solid sorbents significantly. She put forward the two-stage solid sorption freezing cycle, and constructed a new concept of solid sorption heat pipe. She put forward the resorption multi-mode energy conversion cycle. The awards she has gotten on the research work included the Second Prize of National Natural Science Award, China Youth Science and Technology Award, the National Science Fund for Distinguished Young Scholars, EU Marie Curie International Incoming Fellowship, Royal Society International Incoming Fellowship, IIR Young Researchers award, etc.

Guoliang An is the Ph.D. candidates under the supervision of Prof. Liwei Wang in Shanghai Jiao Tong University. His research work focuses on the advanced sorption materials development, kinetic properties and thermodynamics optimization. In the past four years, he has published over 15 journal articles, including Journal of Materials Chemistry A, Chemical Engineering Journal, Renewable and Sustainable Energy Reviews and Energy.

Dr. Jiao Gao graduated from Shanghai Jiao Tong University, and had been supervised by Prof. Liwei Wang. She reasearched on the development of composite sorlid sorbents and their application in refrigeration and energy storage systems.

Ruzhu Wang is a Professor of Institute of Refrigeration and Cryogenics in Shanghai Jiao Tong University. He has engaged in refrigeration and heat pump research for a long time, and has made systematic and creative achievements in the efficient conversion and utilization of low-grade thermal energy. On the research work he has won the Second Prize of National Natural Science Award and National Technological Invention Award, gold medal of International Refrigeration J&E Hall Award, Nukiyama Memorial Award of International Thermal Science, and Asian Refrigeration Academic Award, etc.

Abbreviations

A	Coefficient in Clausius-Clapeyron equation; heat transfer area, m^2
A_0	Dynamic coefficient
A_c	Cross-section area, m^2
A_{cabin}	The surface area of the cabin, m^2
A_{fe}	Anterior factor
c_i	Specific heat of air, kJ/(kg K)
c_o	Specific heat flow rate of heating oil, kJ/(kg K)
C_{am}	Specific heat capacity of ammonia J/(mol K), J/(kg °C)
C_p	Isobaric specific heat, J/(mol K), J/(kg °C)
C_{ps}	The isobaric specific heat of solid material, J/(mol K), J/(kg °C)
C_w	Specific heat of water, J/(kg K)
COP	Coefficient of performance for refrigeration
COP_{ref}	Refrigeration coefficient of performance
D_s, D_{so}	Surface diffusion coefficient, m^2/s
E	Activation energy of the reactants, J/mol
E_a	Activated energy for surface diffusion, sorption activated energy, J/mol
E_d	Desorption activated energy, J/mol
$E_{h,in}$	Heat and refrigeration exergy, J
h_{exh}	The convection heat transfer coefficient of the exhaust gas, W/(K m^2)
h_T	The latent heat of vaporization, J/kg
k	Coefficient in D-R equation
k_s	Mass transfer coefficient inside the solid phase film, kg/(m^2 s)
K	The coefficient for D-R equation, equilibrium constant of the reaction, permeability (m^2)
K_a	Coefficient for the reaction rate in sorption process, 1/(m^2 s)
K_{cabin}	Total heat transfer coefficient, W/(K m^2)
K_d	Coefficient for the reaction rate in desorption process, 1/(m^2 s)
K_i	The dynamic coefficient
K_n	Knudsen diffusion rate
K_s	The total mass transfer coefficient (kg/(m^2 s)), permeability (m^2/s)
K_{sap}	Surface diffusion rate coefficient 1/s

K_{st}	Dynamic coefficient
l	The characteristics diameter of the cross section, mass transfer scale, m
L_a	Latent heat of ammonia, kJ/kg
m_a	Mass of the multi-salt sorbent, kg
m_{me}	The mass of metal, kg
m_o	specific mass flow rate of heating oil, kg/s
$Q_{ins,ref}$	The instantaneous refrigerating capacity, J or kJ
Q_{l-s}	Reaction heat of the solid sorbent and liquid sorbate, J or kJ
r_{as}	Ratio between expansion space and volume of sorbent
r_c	Diameter of reaction surface, m
r_g	Radius of grain, radius of the crystalline grain m
R	The universal gas constant, J/(mol K); thermal resistance, K/W
R_e	Reynolds number, °C/W
R_i	Increase ratio of calcium chloride salt
R_p	Average diameter of the sorbent granules, m
RH	The vapor pressure divided by the saturated vapor pressure
RR	The moisture recovery ratio
S	The exchange surface, the cross-sectional area, m^2
SCP	Specific cooling power per kg sorbent, W/kg
t	Variation of time, s or min
t_{cyc}	Cycle time, s or min
t_{coo}	The cooling time, s or min
$t_{h,in}$	The charging time, s or min
t_r	The reaction time corresponding to a given global extent, s or min
t_{ref}	Refrigerating time, s or min
T	Temperature, K or °C
T_0	The ambient temperature, K or °C
$T_{ambient}$	The ambient temperature, K or °C
T_c	Constrained temperature, condensing temperature, K or °C
$T_{chilled}$	The chilled air temperature, K or °C
T_e	The evaporation temperature, K or °C
m_{so}	The total mass of sorbents, kg
m_{tot}	Total mass of composite sorbent, kg
m_x, m_y	Reaction order
M_{ci}	The mass of dry composite sample, kg
M_{Mi}	The mass of dry host matrix, kg
$M_{tot,ads}$	Total mass of composite sorbent, kg
$M_{tube,met}$	The metallic mass of the unit tube, kg
Ma	Reaction dynamic coefficient for sorption
n	Coefficient in D-A equation, coefficient for reaction equilibrium, reaction order
n_s	Number of moles of salt
N	number of moles of gas sorbed per mole of salt
N_g	Molar solid sorption quantity, mol/mol

p_c	Constrained pressure, Pa
p_{eq}	Equilibrium pressure, Pa
p_i	Pressure of the vapor reactant interface, pressure inside pore, Pa
p_s	The saturated water vapor pressure, Pa
p_v	The vapor pressure, Pa
P	The wet perimeter length, m
q	Heat flux density, W/m²
q_{axi}	the axial heat flux, W/m²
q_c	Heat carried away by the cooling air during sorption process, kW
$q_{h,}$	heat transferred to the sorption bed during desorption process, kW
q_i	Heat transfer rate heat, kW
q_r	Refrigeration capacity, kW
$Q_{ave,des}$	The desorption heat, J or kJ
$Q_{ave,heat}$	The average heat input, J or kJ
$Q_{ave,ref}$	The average refrigerating capacity, J or kJ
$Q_{ave,tube,heat}$	The average heat input of each unit tube in the desorption process, J or kJ
Q_{con}	Condensation heat, J or kJ
Q_{des}	Desorption heat, J or kJ
Q_e	Cooling amount generated in the evaporator, J or kJ
Q_h	Heating capacity, J or kJ
$Q_{h,lat}$	The latent heat output, J or kJ
$Q_{h,rel}$	The discharging heat, J or kJ
$Q_{h,sen}$	The sensible heat output of reactor during cooling phase, J or kJ
Q_{in}	Heat input, J or kJ
$Q_{ins,heat}$	The instantaneous heat input, J or kJ
T_{en}	The environmental temperature, K or °C
T_{eq}	The equivalent temperature for reaction, K or °C
T_{in}	The inlet temperature, K or °C
T_h	The highest heating temperature, relatively high temperature, K or °C
T_l	Relatively low temperature, K or °C
T_{out}	The outlet temperature, K or °C
T_s	Saturated temperature, sorption temperature, K or °C
U	The global exchange coefficient
U_p	The global heat exchange coefficient
v	Fluid velocity, m/s
V_a	Volume of the sorbent, m³
V_{A_B}	The bulk volume of AdBlue, m³
V_{D_O}	The bulk volume of diesel oil, m³
$V_{g,r}$	Radial vapor diffusive velocity, m/s
V_i	Volumetric flow rate of air, m³/h
V_m	The molar volume of ammoniate chlorides, m³/mol; the mass flow rate of dry air, kg/s
V_R	Total volume inside the sorption bed, m³
V_{sx}	Real volume at zero porosity, m³

W	Electrical power generated by turbine, W or kW
W_e	Total rated power of fans, W or kW
W_h	Heating power, W or kW
x	Solid sorption quantity, kg/kg
x^*	The local solid sorption quantity, kg/kg
x_0	Maximum sorption quantity, kg/kg
x_{max}	Maximum sorption quantity, kg/kg
X	Conversion degree
Y	Conversion degree
α	Thermal diffusivity, m^2/s
ε	The ratio of latent heat to sensible heat
ε_a	Porosity of sorbent, kg/m^3
μ	Chemical potential, dynamic viscosity, kg/(m s)
λ_a	Thermal conductivity in the axial direction, W/(m °C)
λ_{eff}	Effective thermal conductivity, W/(m °C)
$\lambda_{M\parallel}$	The limit parallel thermal conductivity, W/(m °C)
$\lambda_{M\perp}$	The limit perpendicular thermal conductivity, W/(m °C)
λ_r	Thermal conductivity in the radial direction, W/(m °C)
λ_{sor}	The thermal conductivity of the composite sorbent, W/(m °C)
ρ_b	Density of the consolidated matrix, kg/m^3
ρ_i	Density of air, kg/m^3
ρ_M	Maximum bulk density, kg/m^3
λ	Thermal conductivity of sorbent
η	Collector efficiency, thermal efficiency of the solar air collector, thermoelectric conversion efficiency
η_{ene}	Total energy efficiency
η_{ene}	Total exergy efficiency
ΔD	The real-time amount of water vapor, kg/s
ΔF	Free sorption energy, J/mol
$\Delta H, \Delta H_r$	Change of the chemical reaction heat, J/mol
ΔH^0	Change of the standard enthalpy, J/mol
$\Delta M_{tube,ref}$	The refrigerant mass that can be sorbed by the composite sorbent in the unit tube, kg
ΔS^0	Change of the standard entropy, J/(mol K)
ΔT	Temperature difference, °C
Δx	Cycle sorption quantity, kg/kg
$\Delta X(r, t_r)$	The local extent of the reaction
ρ	Local bulk density, fluid density, kg/m^3
η_{El}	Theoretical energy coefficient of performance for electrical generation
ν	The stoichiometric coefficient of the chemical reaction, kinematic viscosity, m^2/s
θ	Degree of coverage, solar elevation angle, the heat load friction
$\theta(r, t)$	The local dimensionless temperature
θ^*	The corrected Carnot temperature

φ	Relative humidity
ϕ	The safety factor
a, ad, ads	Sorption, sorbent
am	Ammonia
c	Condensation, cooling
C	Refrigeration, cooling
ca	Composite sorbent
cal	Calculation
cond	Condensation
Cool	Refrigeration, cooling
d, des	Desorption
dil	Dilution
e, eva	Evaporation, refrigeration
ea	Equilibrium sorption
ed	Equilibrium desorption
eff	Effective
eq	Equilibrium
exp	Experiment
f	Fluid
g	Generation, gas
h, H	Heating, highest, heat pump
hp	Heat pipe
Heat	Heating, heat pump
in	Inlet
l	Liquid
m	Cooling media, metal
max	Maximum
mb	The metal back plate
mi	The metal cooling pipe
ref	Refrigerant, refrigeration
reg	Regeneration, heat recovery
s, syn	Synthesisation
s, sat	Saturation
so	Solid
w	Wall
wv	Water vapor

Chapter 1
Introduction

Abstract In this chapter, we give a summary for the solid composite sorbents, including the fundamental principle, theory and working pairs of solid sorption, properties of composite sorbents with expanded natural graphite (ENG) matrix, as well as the energy conversion cycles of solid composite sorbents. Basically, the features and the development of solid composite sorbents are discussed thoroughly.

Keywords Composite sorbent · Expanded natural graphite · Working pair · Cycle · Energy conversion

Sustainable development is the common pursuit of people all over the world, and for which the energy utilization is the key factor. The rapid development of economy is usually accompanied by huge energy consumption, as well as the pollution to the environment. One important strategy for the sustainable development is to coordinate the relations among the energy utilization, economy development and environmental protection.

In terms of environmental protection, the destruction of the ozone layer by chlorofluorocarbons (CFCs) has become a recognized problem all over the world. CFCs are very important substances in compression refrigeration. As an alternative substance, HCFCs can only be used temporarily because they also affect the ozone layer. At the same time, for central heating, the combustion of gas and coal increases carbon dioxide emissions. As people all over the world demand more and more comfortable living conditions, the increasingly serious greenhouse effect and ozone layer depletion have gotten more and more attention. To find a type of green technology for air conditioning, heat pump, energy storage, and electricity generation is very important for solving the problems caused by the conventional technology.

Another key issue for refrigeration and heat pump is the energy utilization. Conventional compression refrigerators and heat pumps are usually driven by electricity. The electricity demand increases fast with the development of the society. According to data provided by the US Department of Energy, air conditioning accounts for about 12% of US home energy expenditures in 2018–2019. Meanwhile, the electricity consumption by the air conditioner in China occupies 34% of the whole world consumption in 2018–2019.The energy efficiency of electricity generation by thermal power plant is only about 40% to 50%, of which a great part of the energy has been released to the environment as the waste heat with the temperature around 70–200 °C. Meanwhile the solar energy and geothermal heat also exist with great amount in the environment as low-grade energy. The development of refrigeration, heat pump and energy storage technologies driven by low-grade thermal energy such as factory waste heat, solar energy, and geothermal energy will be a solution for future energy conservation.

Sorption refrigeration, heat pump, energy storage and water production technologies are driven by low-grade thermal energy and adopt green refrigerants, which are in harmony with the current energy and environmental sustainability requirements. First of all, sorption refrigeration technology has very low requirements for electricity. Secondly, the refrigerants of sorption refrigeration are generally water, ammonia, methanol and other substances, which are green refrigerants with zero ODP (ozone depletion potential) and zero GWP (greenhouse effect potential).

As a kind of sorption technology, the solid sorption has received more and more attention since the 1970s. Compared with absorption refrigeration technology, the solid sorption refrigeration has great scopes of sorbents, including different physical and chemical sorbents, which could satisfy the low-grade heat in the range of 50–400 °C. Secondly, solid sorption refrigeration does not require solution pumps and rectification equipments, so there are no refrigerant pollution and solution crystallization problems that often occur in absorption refrigeration technology. But in general, the efficiency of solid sorption refrigeration is not as good as absorption refrigeration, and the solid sorption bed also has the disadvantage of large size. Solid sorption technology has been recognized as an essential supplementary technology for the absorption technology because of these advantages and disadvantages.

1.1 Solid Sorption Phenomena

According to the different forces for solid sorption processes, the solid sorption is divided into physical solid sorption and solid chemisorption [1]. The physical solid sorption is accomplished by van der Waals forces among the molecules, which generally occur on the surface of the sorbent. The physical solid sorption doesn't have the selectivity, which means the multi-layer solid sorption can be formed. The phenomenon of physical solid sorption can be regarded as the condensation process of the refrigerant inside the sorbent. For most sorbents, the heat of solid sorption is similar to the condensation heat of the refrigerant. The sorbent molecules for the physical solid sorption won't be decomposed during the desorption process. The solid chemisorption is different from the physical solid sorption. In the solid chemisorption process, the sorbent reacts chemically with the sorbate to form new molecules. Usually the sorbate monolayer will react with the chemical sorbent, and after this reaction, the chemical sorbent cannot sorb more molecules. The newly formed molecules will decompose during the desorption process. The solid sorption/desorption heat will be the chemical reaction heat, which is much larger than the physical solid sorption heat. The solid chemisorption has the selectivity. For instance, H_2 can be sorbed by W, Pt, and Ni, but cannot be sorbed by Cu, Ag, or Zn. Because the effective distance of Van der Waals force is inversely proportional to the power of 7 of the distance and is much larger than the effective distance of chemical reaction, it is recognized by the academics that physical solid sorption occurs before chemical reaction. Therefore, when the sorbate molecule approaches to the solid sorbent, the physical solid sorption will proceed first, and when the distance decreases to the threshold level, it will change to solid chemisorption process.

The physical solid sorption/desorption mainly depends on the heat and mass transfer performance of the sorbent. For the desorption process, due to the high pressure the heat transfer performance will be the main indicator to measure the system performance. If the heat transfer performance is enhanced, the main problem of the solid sorption system will be the gas permeability inside the sorbent. Generally, the smaller the sorbent particles, the higher the permeability. The kinetic reaction rate also affects the solid sorption/desorption rate.

Since the chemical reaction occurs in the sorption process, the solid chemisorption will be affected by the heat and mass transfer process of the sorbent as well as the chemical reaction kinetics. At the same time, since the desorption activation energy is the sum of the solid sorption activation energy and the solid sorption heat, the desorption activation energy is always far greater than the solid sorption activation energy. This phenomenon will cause a serious hysteresis between solid sorption and desorption [2].

For solid sorption refrigeration, most refrigerant molecules are polar molecular gases, which can be sorbed under van der Waals forces. For example, ammonia, methanol and hydrocarbons can be sorbed by activated carbon, zeolite and silica gel, respectively. The cycle solid sorption capacity of physical solid sorption is generally 10–20%, while the solid chemisorption has larger cycle sorption capacity than that of physical solid sorption. For instance, the cycle solid sorption capacity for $CaCl_2$ is always larger than 0.4 g/g.

The advantage of solid chemisorption refrigeration is the larger solid sorption/desorption capacity, which is essential to increase the specific cooling capacity per kilogram of sorbent (SCP). However, expansion and agglomeration will occur during the sorption process of chemical sorbents. In order to ensure mass transfer performance, the expansion space should generally be maintained at the level of twice the volume of the sorbent. Solidified composite sorbents have been developed, and the porous thermal conductive matrix was used to maintain the reasonable permeability of the sorbent and to improve the heat transfer performance, thereby significantly increase the volume filling capacity and volume cooling capacity.

1.2 Fundamental Principle of Solid Sorption

The basic principle of solid sorption is demonstrated by the solid chemisorption refrigeration cycle in Fig. 1.1, and the relative thermodynamic cycle is shown in Fig. 1.2.

As shown in Fig. 1.1, the chemisorption refrigeration cycle consists of a sorption bed, an evaporator, a condenser and a liquid storage tank. When the sorption bed is cooled by outside air, the pressure inside the sorption bed decreases, and the refrigerant inside the liquid storage tank flows into the evaporator, evaporates under the pressure difference between sorption bed and evaporator, and is sorbed by the sorbent inside the sorption bed. The evaporation process of refrigerant produces cooling power. When the sorption is saturated, the refrigeration will stop. The pressure of the sorption bed increases when it is heated by the low-grade thermal energy. The refrigerant in the sorbent bed is desorbed under the pressure difference between

Fig. 1.1 The chemisorption refrigeration system for refrigerated truck

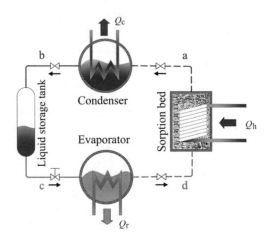

Fig. 1.2 The p–T diagram of chemisorption refrigeration cycle

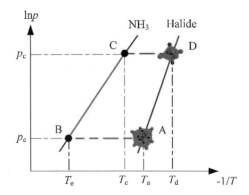

the sorbent bed and the condenser, and then condenses in the condenser which is cooled by the ambient air.

The whole process can be summarized in detail by analyzing the halide-ammonia working pair in Fig. 1.2 as follows:

(1) Desorption stage: The sorption bed is heated at this stage. When the temperature rises to T_d, ammonia will be desorbed from the sorption bed. The desorbed ammonia vapor flows into the condenser (line DC in Fig. 1.2), and then condenses in the condenser (point C in Fig. 1.2).

(2) Sorption stage: The sorption bed is cooled during this stage. The pressure drops sharply when liquid ammonia enters the evaporator after passing through the expansion valve (line CB in Fig. 1.2). In the evaporator, ammonia vapor flows back to the sorption bed (curve BA in Fig. 1.2) to complete the cycle. The evaporation of ammonia provides cooling power.

As mentioned above, the simple conventional cycle is an intermittent refrigeration cycle. For continuous refrigeration output, a solid sorption system with two or more sorbent beds needs to be designed. In this case, the heating and cooling processes of multiple sorption beds can complement each other.

1.3 Solid Sorption Working Pairs

Solid sorption working pairs are important for the performance of refrigeration and heat pump systems. The working pairs have great influence on the coefficient of performance (COP) of systems, the temperature rising rate of the sorbent bed and the initial investment. In order to obtain high refrigeration capacity, it is necessary to select a suitable solid sorption working pair according to the temperature of heat source, and select a suitable solid sorption refrigeration cycle according to actual requirements. Different solid sorption refrigerant pairs have different application ranges and performance. The common solid sorption refrigeration working pairs mainly include: activated carbon-methanol, activated carbon fiber-methanol, activated carbon-ammonia, zeolite-water, silica gel-water, metal hydrides-hydrogen, calcium chloride-ammonia, and strontium chloride-ammonia etc. [3]. Recently, studies also show that composite solid sorbents are effective to enhance heat and mass transfer performance, which are promising for refrigeration technology [4–6].

For working pairs of physical solid sorption, carbon-methanol working pair has a large solid sorption and desorption concentration. Its desorption temperature is around 100 °C and it has the advantage of low solid sorption heat, which is around 1800–2000 kJ/kg. Methanol refrigerant can be applied to make ice because its freezing point is below 0 °C, and its highest desorption temperature cannot exceed 120 °C, otherwise methanol will decompose. The advantage of activated carbon-ammonia system is the low evaporation temperature of the refrigerant which is commonly used for making ice. Meanwhile, ammonia can also adapt to high heat source temperature. For silica gel-water working pair, the desorption temperature cannot be too high. If the temperature is higher than 120 °C, the silica gel will be destroyed. Therefore, it is commonly used for low-temperature heat source. The zeolite-water working pair has a wide desorption temperature range (70–250 °C). The solid sorption heat is about 3200–4200 kJ/kg, and the evaporation latent heat of water is 2400–2600 kJ/kg. Zeolite-water is quite stable and will not be destroyed at high temperature. However, its disadvantage is that the solid sorption heat is too high, which leads to low COP. The evaporation temperature needs to be higher than 0 °C, which cannot be used for making ice. In addition, the system is a vacuum system, which has higher requirements for vacuum sealing. Furthermore, the low evaporation pressure causes slow solid sorption process than other working pairs.

The chemisorption working pairs mainly include hydride-hydrogen, metal chloride (salt)-ammonia, metal oxide–water, and metal oxide–carbon dioxide. The metal hydride-hydrogen system uses solid sorption process and desorption process between metal or alloy and hydrogen for refrigeration. It is characterized by large sorption/desorption heat, especially suitable for advanced porous metal hydride (PMH) or, Misch metal (Mm) alloy matrix, including Ni, Fe, La, Al. This type of working pair is usually used for solid sorption heat pumps due to its higher sorption heat and higher sorption concentration. The metal chloride-ammonia working pair has a large sorption capacity. For example, for the calcium chloride-ammonia working pair, 1 mol of calcium chloride can sorb 8 mol of ammonia. At the same time, the boiling point of ammonia is lower than −34 °C, which can be used to make ice. Meanwhile, the refrigerator works under positive pressure, such a feature is more reliable for less risk of leakage than vacuum conditions. Metal oxide–water and metal oxide–carbon dioxide have the advantages of high energy storage during hydration and carbonization [7, 8]. Taking calcium oxide as an example, the energy storage during the hydration and carbonization process is 800–900 kJ/kg, which makes it possible to develop efficient heat pump systems.

However, chemisorption has the feature of agglomeration and expansion, which will lead to the problems of low permeability and poor heat transfer performance. In order to overcome this problem, a porous thermal conductive matrix is proposed to improve the heat and mass transfer performance of chemical sorbents. The research on this type of sorbent mainly focuses on the composite sorbent based on ENG, activated carbon fiber and activated carbon. Studies have shown that this composite can increase the volumetric filling capacity and volumetric cooling capacity of the sorbent [4–6].

1.4 Solid Composite Sorbents with ENG Matrix

ENG has been early invented by the Carburet Company in US [9]. It is prepared from expandable graphite by electrochemical and chemical oxidation methods [10]. Generally, it contains abundant multi-pores ranging from 20 to 50 nm. ENG maintains a layered structure similar to that of natural graphite sheets, but the interlayer spacing is larger [11], which ranges from 0.335 nm to about 0.8 nm [12, 13]. Due to its excellent properties such as compatibility, flexibility, anti-corrosion, and so on, it has been widely used in sealing, catalysis, aerospace, military, environmental protection and other fields as a kind of functional carbon material [14].

Among all the advantages, the high thermal conductivity after compression is the evident merit of ENG, so it has been used extensively on the phase change materials (PCMs) for energy storage [15–17]. PCMs can store a large amount of thermal energy with a constant temperature due to their high latent heat during the phase transition, which offers an effective method for the appropriate utilization of low-grade heat [18, 19]. However, the applications of PCMs in these thermal energy storage systems are limited by their relatively low thermal conductivity [20], which in turn reduces the rate of thermal energy storage and extraction during melting and solidification processes [21, 22]. Several techniques for enhancing the heat transfer have been tested experimentally [23–26], among which ENG is proved being a prospective additive [27], and academics found that the addition of 10 wt.% ENG leads to about five-fold increase in thermal conductivity [28]. PCM and consolidated ENG (CENG) composites applied for paraffin's [29] for low temperature thermal energy storage (below 100 °C) and inorganic salts [30] for high temperature range (300 °C) have been proved with high energetic properties, and the corresponding thermal conductivity can range from 5 to 70 W/(m K) [29–31]. Moreover, the combination of CENG and latent heat storage materials leads to black materials which could potentially have a high absorptivity for the solar spectra [32]. Experiments indicated that the mixture of metal foam and ENG could improve the heating rate of solid $NaNO_3$ by 190% and 250%, respectively [33]. The optimal percentage of ENG can achieve the uniform distribution of PCMs inside and can effectively prevent the liquid leakage [34–36], and consequently forms the stable composite PCM [15, 37]. Investigation also indicated that the composite of PCM/ENG demonstrates excellent thermal cycling stability [38–40].

Early study on consolidated solid sorbent with ENG researched by Mauran et al. revealed the excellent heat transfer property [41]. They impregnated $CaCl_2$ in ENG that is termed IMPEX (IMPregnated blocks of recompressed expanded graphite) for solid sorption process of ammonia [41, 42], patented and manufactured by La Carbone Lorraine. It has strong anisotropic heat and mass transfer characteristics. The layered structure of CENG improves gas diffusion perpendicular to the compression direction [16]. After that, the ENG is introduced to coated beds that insert solid sorbents into the ENG plates [43]. For this coated bed, the contact between the metal wall and the solid sorbent is not as tight as the coated tube, but due to particle sorption the diameter of the agent is only a few microns, so the heat capacity ratio of the sorbent is greatly improved. Because the mass transfer performance is another important index for evaluating the property of solid sorbents, Han and Lee studied the permeability of composites of ENG and metallic salt for heat pumps, and found that the permeability of gas is in the range of 10^{-16}–10^{-12} m^2, depending on the working pair, bulk density and mass fraction of the ENG [44]. Fujioka et al. developed a composite sorbent composed of $CaCl_2$, ENG, and AC fibre, and they improved the thermal conductivity of the sorbents from less than 0.15 to over 0.6 W/(m K) [45]. Consolidated composites of $LaNi_5$ and CENG improve the thermal conductivity of the metal hydride from 0.1 W/(m K) to 3–6 W/(m K) showing a good potential for developing devices for hydrogen storage with high sorption density [46].

Normally the ENG is produced by expanding the granular natural graphite in oven at 600–800 °C, and by such a technique the thermal conductivity is limited. For example, a pure pellet of ENG with a porosity of 79.1% only has a thermal conductivity of 8 W/(m K) when the density of CENG is 1250 kg/m^3 [47]. As a result, for the consolidated composite ENG/sorbent beds the thermal conductivity is always lower than 6–7 W/(m K) [48, 49]. Py et al. proposed a new method developing the composite solid sorbents [50] that is a little complex, which required the precursor and the activation agent for composite preparation of AC/ENG, followed by compression and activation (physical or chemical). Such methods achieve a thermal conductivity as high as 32 W/(m K) [50], and the composites are evaluated for CO_2 sorption for gas purification [48].

Chemical and electrochemical oxidation of graphite in acid leads to the formation of salt-like intercalated compounds [51–53], i.e. forming graphite intercalation compound (GIC). One type of ENG-GIC evaluated as an enhancer of thermal conductivity for PCMs was produced by consolidating graphite flakes that have been impregnated with sulphuric and nitric acid and heated at 900 °C [54]. The results show that the thermal conductivity of the consolidated material is as high as 16.6 W/(m K) [55] when the density of the graphite is 220 kg/m^3. Such materials are also successfully utilized for the improvement of performance in PEM fuel cells [56] or batteries [57] and solid sorbents [58].

1.5   Solid Sorption Theory

The research on solid sorption refrigeration develops from equilibrium solid sorption, which has equilibrium pressure corresponding to the sorption temperature. With the development of enhanced heat and mass transfer technology, the non-equilibrium status is more and more important for kinetic characteristics of solid sorption/desorption.

For the physical solid sorption, Critoph proposed a simplified form of the D-A equation [59, 60], in which only the temperature is considered. It is an empirical equation widely used to evaluate the solid sorption performance under equilibrium conditions, but it cannot be used to analyze the solid sorption performance of non-equilibrium. For this reason, Sokoda established a model of the solid sorption rate for which the dynamic process of solid sorption is considered as the mass transfer process of the gases inside the solid sorption systems:

$$\frac{\mathrm{d}x}{\mathrm{d}t} = K_s a_p (x^* - x) \tag{1.1}$$

$$K_s a_p = \frac{15 D_{so}}{R_p{}^2} \exp\left(\frac{-E_a}{RT}\right) \tag{1.2}$$

where x^* is the local solid sorption capacity, $K_s a_p$ is the coefficient for the rate of surface diffusion, D_{so} is the surface diffusion coefficient, E_a is activated energy for the surface diffusion, R_p is the average diameter of the sorbent granules. This equation is mainly for the silica gel-water working pair. This model can be utilized for other working pairs but it needs the amendment of the coefficients in the equations. For example, Passos [61] revised the coefficients of this equation for researching the activated carbon-methanol working pair.

Compared with physical solid sorption, the solid chemisorption model is very complicated. There are three main types of analysis models: local models, global models and analytical models. The local model takes the mass and heat transfer, as well as the dynamics of the small volume into account, resulting in a partial derivative equation and requiring numerically solved. The global model considers the variables and average values of reactor characteristics, such as permeability, thermal conductivity, heat capacity, etc., for simulation. The numerical solution of the global model gives a set of differential equations. The analytical model considers the average of the variables within the reaction time, and these differential equations are only related to spatial variables. Spinner and Rhaault [62] studied the non-uniform kinetics based on the study of dynamic solid sorption rate. Then, based on the research results of Spinner and Rheault, Mazet [63] and Lebrun [64] revised the equations proposed by Tykodi [65] and Flanagan [66]:

$$\frac{dx}{dt} = K_i (1 - x) \exp(-A_0/T) \ln\left(\frac{p_c}{p_{eq}(T)}\right) \tag{1.3}$$

where x is solid sorption capacity, dx/dt is solid sorption rate; K_i is dynamic coefficient, and subscript $i = s$ for solid sorption process, $i = d$ for desorption process. p_c is the constrained pressure of condenser and evaporator, p_{eq} is equilibrium pressure, and T is solid sorption temperature.

Due to the small influence of A_0 in the experiment, Mazet performed a logarithmic transformation of Eq. 1.3 [63]. Based on this, Goetz [67] established a model that considers the internal mass transfer performance of particles, which is

$$\frac{dN_g}{dt} = 4\pi r_c^2 K_i \left(\frac{p_c - p_{eq}(T)}{p_{eq}(T)}\right)^{Ma} \tag{1.4}$$

where N_g is molar solid sorption capacity, r_c is the diameter of reaction surface, and Ma is the reaction dynamic coefficient.

Another formula [68] for the reaction rate which considers the Darcy equation for reaction surface and grain surface is

$$\frac{dx}{dt} = f(x, r_g) \left(\frac{p_c - p_i}{T_c}\right) K_n(m, c) \tag{1.5}$$

where K_n is Knudsen diffusion rate that is related with the diameter of pore and porosity, $f(x,r_g)$ is a function which is related with solid sorption capacity x and the radius of grain r_g. p_i is the pressure inside pore.

Usually, the solid sorption model is also used in the desorption process. However, Mazet pointed out that there is a quasi-equilibrium zone in the solid–gas reaction [69], and Goetz [67] also considered this zone in his research work. Shanghai Jiao Tong University (SJTU) studied the chemisorption and composite solid sorption under non-equilibrium heating and cooling conditions, and the results showed that there is a serious hysteresis in the solid sorption and desorption processes [70]. The actual refrigeration process is always in a non-equilibrium state, so the chemisorption model needs to consider this hysteresis.

Another problem of solid chemisorption is the difference between solid chemisorption model and composite solid sorption model. The main part in the composite sorbent is the chemical sorbent, so the chemical sorbent model is generally used for simulating the composite solid sorption system. This simulation is acceptable for the equilibrium process. For non-equilibrium processes, composite solid sorption is complicated because it includes heat and mass transfer processes in chemical sorbents and porous media. During the reaction stage, the volume of the chemical sorbent and the density of the porous additive will change, which will affect the solid sorption performance. Therefore, for non-equilibrium solid sorption model, the heat and mass transfer performance of chemical and porous materials needs to be considered.

1.6 Energy Conversion Cycles of Solid Composite Sorbents

As a type of energy saving and environmental benign technology, the solid sorption has been applied in many energy conversion cycles, including refrigeration [71], water production [72], eliminating NO_x emission [73], energy storage [74], and electricity cogeneration [75]. The solid sorption technology has been paid more and more attention because of its advantages in many aspects. For example, through the energy and exergy analysis of various studies, the sorption energy storage technology has been proven to be one of the most effective thermal energy storage (TES) technologies, and it is also considered to be able to meet the requirements of seasonal energy storage, with negelible energy loss.

So in Chaps. 5 and 6, different types of energy conversion cycles will be introduced in detail to show the advantages and characteristics of using the solid composite sorbents.

In summary, as an energy-saving and environmentally friendly technology, solid sorption has attracted more and more attention. Researchers continue to work hard and have achieved many results, laying a good foundation for further development. But the wide application of this technology still has a long way to go. The following chapters will give a detailed summary and analysis of the achievements and existing problems in the research work.

References

1. Gao J, Wang LW, Wang RZ et al (2017) Solution to the sorption hysteresis by novel compact composite multi-salt sorbents. Appl Therm Eng 111:580–585
2. Wang, LW, Wang RZ, Wu JY, Wang K (2004) Solid sorption performances and refrigeration application of solid sorption working pair of $CaCl_2$–NH_3. Sci China Ser E 47(2):173–85
3. Wang LW, Wang RZ, Oliverira RG (2009) A review on adsorption working pairs for refrigeration. Renew Sustain Energy Rev 13:518–534
4. Fayazmanesh K, Salari S, Bahrami M (2017) Effective thermal conductivity modeling of consolidated sorption composites containing graphite flakes. Int J Heat Mass Transf 115:73–79
5. Tanashev YY, Krainov AV, Aristov YJ (2013) Thermal conductivity of composite sorbents "salts in porous matrix" for heat storage and transformation. Appl Therm Eng 61:401–407
6. Wang LW, Metcalf SJ, Critoph RE et al (2011) Thermal conductivity and permeability of consolidated expanded natural graphite treated with sulphuric acid. Carbon 49:4812–4819
7. Al-Shankiti IA, Ehrhart BD, Ward BJ et al (2019) Particle design and oxidation kinetics of iron-manganese oxide redox materials for thermochemical energy storage. Sol Energy 183:17–29
8. Prieto C, Cooper P, Fernández AI et al (2016) Review of technology: thermochemical energy storage for concentrated solar power plants. Renew Sustain Energy Rev 60:909–929
9. Li JH, Feng LL, Jia ZX (2006) Preparation of expanded graphite with 160 μm mesh of fine flake graphite. Mater Lett 60:746–749
10. Li JH, Da HF, Liu Q, Liu SF (2006) Preparation of sulfur-free expanded graphite with 320 μm mesh of flake graphite. Mater Lett 60:3927–3930
11. Celzard A, Mareche JF, Furdin G, Puricelli S (2000) Electrical conductivity of anisotropic expanded graphite-based monoliths. J Phys D Appl Phys 33:3094–3101
12. Kang F, Zhang TY, Leng Y (1997) Electrochemical behavior graphite in electrolyte of sulfuric and acetic acid. Carbon 35:1167–1173
13. Chen GH, Wu DJ, Weng WC, He B, Yan WL (2001) Preparation of polystyrene-graphite conducting nanocomposites via intercalation polymerization. Polym Int 50(9):980–985
14. Kang WS, Rhee KY, Park SJ (2016) Thermal impact and toughness behaviors of expanded graphite/graphite oxide-filled epoxy composites. Compos B Eng 94:238–244
15. Koukou MK, Vrachopoulos MG, Tachos NS et al (2018) Experimental and computational investigation of a latent heat energy storage system with a staggered heat exchanger for various phase change materials. Thermal Sci Eng Progress 7:87–98
16. Mallow A, Abdelaziz O, Graham JS (2016) Thermal charging study of compressed expanded natural graphite/phase change material composites. Carbon 109:495–504
17. Santhi Rekha SM, Sukchai S (2018) Design of phase change material based domestic solar cooking system for both indoor and outdoor cooking applications. J Sol Energy Eng 140:135–142
18. Da Cunha JP, Eames P (2016) Thermal energy storage for low and medium temperature applications using phase change materials-a review. Appl Energy 177:227–238
19. Mohamed SA, Al-Sulaiman FA, Ibrahim NI et al (2017) A review on current status and challenges of inorganic phase change materials for thermal energy storage systems. Renew Sustain Energy Rev 70:1072–1089
20. Zhong LM, Zhang XW, Luan Y, Wang G, Feng YH, Feng DL (2014) Preparation and thermal properties of porous heterogeneous composite phase change materials based on molten salts/expanded graphite. Sol Energy 107:63–73
21. Liu M, Saman W, Bruno F (2012) Development of a novel refrigeration system for refrigerated trucks incorporating phase change material. Appl Energy 92:336–342
22. Elias CN, Stathopoulos VN (2019) A comprehensive review of recent advances in materials aspects of phase change materials in thermal energy storage. Energy Procedia 161:385–394
23. Karaipekli A, Sarı A (2016) Development and thermal performance of pumice/organic PCM/gypsum composite plasters for thermal energy storage in buildings. Sol Energy Mater Sol Cells 149:19–28

24. Merlin K, Delaunay D, Soto J et al (2016) Heat transfer enhancement in latent heat thermal storage systems: comparative study of different solutions and thermal contact investigation between the exchanger and the PCM. Appl Energy 166:107–116
25. Ibrahim NI, Al-Sulaiman FA, Rahman S et al (2017) Heat transfer enhancement of phase change materials for thermal energy storage applications: a critical review. Renew Sustain Energy Rev 74:26–50
26. Chen C, Zhang H, Gao X et al (2016) Numerical and experimental investigation on latent thermal energy storage system with spiral coil tube and paraffin/expanded graphite composite PCM. Energy Convers Manage 126:889–897
27. Wang XL, Guo QG, Zhong YJ, Wei XH, Liu L (2013) Heat transfer enhancement of neopentyl glycol using compressed expanded natural graphite for thermal energy storage. Renew Energy 51:241–246
28. Xia L, Zhang P (2011) Thermal property measurement and heat transfer analysis of acetamide and acetamide/expanded graphite composite phase change material for solar heat storage. Sol Energy Mater Sol Cells 95:2246–2254
29. Py X, Olives R, Mauran S (2001) Paraffin/porous-graphite-matrix composite as a high and constant power thermal storage materials. Int J Heat & Mass Transfer 44:2727–2737
30. Pincemin S, Olives R, Py X, Christ M (2008) Highly conductive composites made of phase change materials and graphite for thermal storage. Sol Energy Mater Sol Cells 92:603–613
31. Zhao YJ, Wang RZ, Wang LW, Yu N (2014) Development of highly conductive $KNO_3/NaNO_3$ composite for TES (thermal energy storage). Energy 70:272–277
32. Haillot D, Goetz V, Py X, Benabdelkarim M (2011) High performance storage composite for the enhancement of solar domestic hot water systems, Part 1: Storage material investigation. Sol Energy 85:1021–1027
33. Wu ZG, Zhao CY (2011) Experimental investigations of porous materials in high temperature thermal energy storage systems. Sol Energy 85:1371–1380
34. Wang SP, Qin P, Fang XM, Zhang ZG, Wang SF, Liu XH (2014) A novel sebacic acid/expanded graphite composite phase change material for solar thermal medium-temperature applications. Sol Energy 99:283–290
35. Xiao M, Feng B, Gong KC (2002) Preparation and performance of shape stabilized phase change thermal storage materials with high thermal conductivity. Energy Conv Manag 43:103–108
36. Cheng WL, Zhang RM, Xie K, Liu N, Wang J (2010) Heat conduction enhanced shape-stabilized paraffin/HDPE composite PCMs by graphite addition: preparation and thermal properties. Sol Energy Mater Sol Cells 94:1636–1642
37. Senthil R, Cheralathan M (2016) Natural heat transfer enhancement methods in phase change material based thermal energy storage. Int J Chem Tech Res 9(5):563–570
38. Lee SY, Shin HK, Park M, Rhee KY, Park SJ (2014) Thermal characterization of rythritol/expanded graphite composites for high thermal storage capacity. Carbon 68:67–72
39. Jadal M, Delaunay D, Luo L, et al (2018) Hybrid analytical and finite element simulation of a latent heat storage exchanger: Comparison of heat transfer models for enhanced thermal conductivity phase change material. International Heat Transfer Conference Digital Library. Begel House Inc
40. Sar A, Bicer A, Karaipekli A, Alkan C, Karadag A (2010) Synthesis, thermal energy storage properties and thermal reliability of some fatty acid esters with glycerol as novel solid-liquid phase change materials. Sol Energy Mater Sol Cells 94:1711–1715
41. Mauran S, Prades P, Haridon FL (1993) Heat and mass transfer in consolidated reaction beds for thermochemical systems. Heat Reco Syst & CHP 13:315–319
42. Rivero-Pacho AM, Critoph RE, Metcalf SJ et al (2015) Ammonia carbon adsorption cycle research at the University of Warwick. Polska Energetyka Słoneczna 1–4:47–54
43. Pons M, Dantzer P (1994) Heat transfer in hydride packed beds. Int J Res Phys Chem & Chem Phys 183:1249–1259
44. Han JH, Lee KH (2001) Gas permeability of expanded graphite-metallic salt composite. Appl Therm Eng 21:453–463

45. Fujioka K, Hatanaka K, Hirata Y (2008) Composite reactants of calcium chloride combined with functional carbon materials for chemical heat pumps. Appl Therm Eng 28:304–310

46. Kim KJ, Montoya B, Razani A, Lee KH (2001) Metal hydride compacts of improved thermal conductivity. Int J Hydrogen Energy 26:609–613

47. Klein HP, Groll M (2004) Heat transfer characteristics of expanded graphite matrices in metal hydride beds. Int J Hydrogen Energy 29:1503–1511

48. Menard D, Py X, Mazet N (2005) Activated carbon monolith of high thermal conductivity for adsorption processes improvement, Part A: Adsorption step. Chem Eng Process 44:1029–1038

49. Ahmet S, Ali K (2007) Thermal conductivity and latent heat thermal energy storage characteristics of paraffin/expanded graphite composite as phase change material. Appl Therm Eng 27:1271–1277

50. Py X, Daguerre E, Menard D (2002) Composites of expanded natural graphite and in situ prepared activated carbons. Carbon 40:1255–1265

51. Inagaki M, Iwashita N, Kouno E (1990) Potential change with intercalation of sulphuric acid into graphite by chemical oxidation. Carbon 28:49–55

52. Metrot A (1981) Charge transfer reactions during anodic oxidation of graphite in H_2SO_4. Synthetic Met 3:201–207

53. Bottomley MJ, Parry GS, Ubbelohde AR, Young DA (1963) Electrochemical preparation of salts from well-oriented graphite. J Chem Soc 1083:5674–5680

54. Wang LW, Metcalf SJ, Thorpe R, Critoph RE, Tamainot-Telto Z (2011) Thermal conductivity and permeability of consolidated expanded natural graphite treated with sulphuric acid. Carbon 49(14):4812–4819

55. Mills A, Farid M, Selman JR, Al-Hallaj S (2006) Thermal conductivity enhancement of phase change materials using a graphite matrix. Appl Therm Eng 26:1652–1661

56. Dhakate SR, Sharm S, Borah M, Mathur RB, Dhami TL (2008) Expanded graphite-based electrically conductive composites as bipolar plate for PEM fuel cell. Int J Hydrogen Energ 33:7146–7152

57. Zhao HP, Ren J, He XM, Li JJ, Jiang CY, Wan CR (2008) Modification of natural graphite for lithium ion batteries. Solid State Sci 10:612–617

58. Wang LW, Metcalf SJ, Thorpe R, Critoph RE, Tamainot-Telto Z (2012) Development of thermal conductive consolidated activated carbon for adsorption refrigeration. Carbon 50:977–986

59. Critoph RE (1988) Performance limitations of solid sorption cycles for solar cooling. Sol Energy 41(1):21–31

60. Tamainot-Telto Z, Critoph RE (1997) Solid sorption refrigerator using monolithic carbon-ammonia pair. Int J Refrig 20(2):146–155

61. Passos EF, Escobedo JF, Meunier F (1989) Simulation of an intermittent adsorptive solar cooling system. Sol Energy 42(2):103–111

62. Spinner B, Rheault F (1985) Kinetics models in solid/gas reactions under imposed pressure and temperature constraints. In: The proceedings of international workshop on heat transformation and storage. ISPRA pp 9–11

63. Mazet N, Amouroux M, Spinner B (1991) Analysis and experimental study of the transformation of non-isothermal solid/gas reacting medium. Chem Eng Commun 99:155–174

64. Lebrun M (1990) Models of heat and mass transfers in solid/gas reactor used as chemical heat pumps. Chem Eng Sci 45:1743–1753

65. Tykodi RJ (1979) Thermodynamics of steady state resistance change transitions in steady-state systems. Bull Chem Soc Japan 552(2):564–568

66. Flanagan TB (1978) Hydrides for energy storage. Oxford Press, England

67. Goetz V, Marty A (1992) A model for reversible solid/gas reactions submitted to temperature and pressure constraints; simulation of the rate of reaction in solid gas reactors used in chemical heat pump. Chem Eng Sci 47(17–18):4445–4454

68. Neveu P, Castaing-Lasvignottes J (1997) Development of a numerical sizing tool for a solid-gas thermochemical transformer-I impact of the microscopic process on the dynamic behaviour of a solid-gas reactor. Appl Therm Eng 17(6):501–518

69. Mazet N, Amouroux M, Spinner B (1991) Analysis and experimental study of the transformation of a non-isothermal solid/gas reaction medium. Chem Eng Commun 99:155–174
70. Zhou ZS, Wang LW, Jiang L, Gao P, Wang RZ (2015) Non-equilibrium sorption performances for composite sorbents of chlorides-ammonia working pairs for refrigeration. Int J Refrig 65:60–68
71. Korhammer K, Neumann K, Opel O et al (2018) Thermodynamic and kinetic study of $CaCl_2$–CH_3OH adducts for solid sorption refrigeration by TGA/DSC. Appl Energy 230:1255–1278
72. Wang JY, Wang RZ, Tu YD et al (2018) Universal scalable sorption-based atmosphere water harvesting. Energy 165:387–395
73. Wang ZX, Wang LW, Gao P et al (2018) Analysis of composite sorbents for ammonia storage to eliminate NOx emission at low temperatures. Appl Therm Eng 128:1382–1390
74. Li TX, Wu S, Yan T et al (2016) A novel solid-gas thermochemical multilevel sorption thermal battery for cascaded solar thermal energy storage. Appl Energy 161:1–10
75. Gao P, Wang L, Wang RZ et al (2017) Simulation and experiments on a solid sorption combined cooling and power system driven by the exhaust waste heat. Front Energy 11:516–526

Chapter 2
Development of Solid Composite Sorbents

Abstract This chapter introduces techniques for developing different solid composite sorbents in detail, including processing ENG without or with graphite intercalation compounds (GICs), and developing the composite solid sorbents with ENG, activated carbon, activated carbon fiber and silica gel by simple mixture and consolidation or impregnation and compression.

Keywords Composite solid sorbent · Expanded natural graphite · Activated carbon · Activated carbon fiber · Silica gel · Consolidation · Impregnation

2.1 Techniques for Processing Natural Graphite

Graphite (Fig. 2.1a) is the stable allotrope of carbon, whose structure is based on parallel layers (Fig. 2.1b) where the atoms form hexagonal cells through covalent bonds 1.42 Å long. Along the c-axis of the structure, Van der Waals force maintains the carbon sheets 3.35 Å apart [1]. As a result, the bond strength is anisotropic and much higher along with the layers than perpendicular to them. Such a feature leads to its effortlessly natural cleaving, and the primary morphology of its particles, i.e., flakes and platelets (Fig. 2.1b). Consequently, it can be expanded mostly in the heating phase (Fig. 2.1c). During the rapid heating process of expandable graphite, the natural graphite flakes are expanded violently along the c-axis direction, and expanded natural graphite (ENG) is formed [2].

Also, due to the anisotropic bond strength, insert molecules and atoms between the carbon layers of natural graphite is easy to achieve. Many graphite intercalation compounds (GICs) are formed by inserting various chemical species between the layers in a graphite host material [3]. Upon rapid heating (usually up to 600–1000 °C), the intercalated molecules are suddenly volatilized or even decomposed with vast flakes expansion. In the heating process, the GIC induces the vaporization of the intercalated species, and hence a significant expansion of the material along the crystallographic c-axis occurs. Exfoliation may be reversible or not, since it depends on both the maximum temperature reached and the heating rate [4]. Irreversibly

© Science Press 2021
L. Wang et al., *Property and Energy Conversion Technology of Solid Composite Sorbents*, Engineering Materials,
https://doi.org/10.1007/978-981-33-6088-4_2

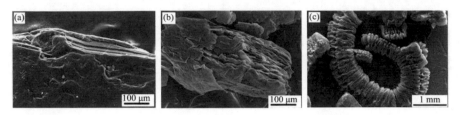

Fig. 2.1 Scanning electron microscope (SEM) images of: **a** natural graphite; **b** expandable graphite; **c** ENG [2]

exfoliated graphite is usually much-expanded, leading to volumetric expansion as high as 300 [5].

2.1.1 ENG Without GIC

The expansion degree of ENG is associated with the duration of expandable graphite and heating temperature. The worm-like ENG (Fig. 2.1c) has a large number of pores, inside which the loose structures are embedded [6].

As the matrix of the solid sorbents, the expansion technique associates with the improvement of permeability and thermal conductivity. To measure the expansion degree, the density is a good indicator since both over expansion and incomplete expansion will lead to larger density, i.e., the optimal expansion technique can make the sample have the smallest density. Under the condition of different duration time and expansion temperature, the density of various samples is shown in Fig. 2.2. For the expansion temperature of 300 °C (Fig. 2.2a), the density decreases when the duration time increases, and the smallest density is gotten when the duration is

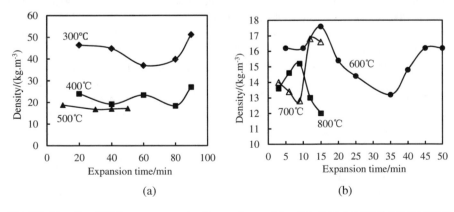

Fig. 2.2 Density of ENG versus expansion temperature and duration time: **a** expansion temperature of 300, 400 and 500 °C; **b** expansion temperature of 600, 700 and 800 °C [7]

Fig. 2.3 Thermal conductivity of ENG versus expansion time at 600, 700, and 800 °C [7]

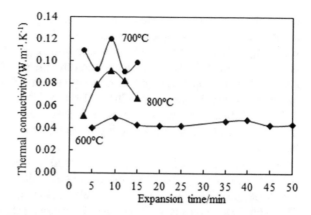

60 min. After that, the density increases since the graphite is over expanded. For the temperature of 400 °C and 500 °C, the smallest density is gotten at about 40 min and 30 min, respectively. The data gotten by the expansion temperature of 600, 700, and 800 °C (Fig. 2.2b) are better than those of 300–500 °C.

The thermal conductivity of various samples under the optimal expansion temperature of 600, 700, and 800 °C is shown in Fig. 2.3. The optimal thermal conductivity is gotten when the expansion time is about 10–15 min for various expansion temperatures, and the highest value is gotten with the expansion temperature of 700 °C. Because the ENG needs to be compressed for producing consolidated sorbents, Tian et al. tested various samples after they are compressed to the same density of 18.8 kg/m^3 [7]. They found that the sample expanded under the temperature of 600 °C and 700 °C has the thermal conductivity of 0.095 and 0.082 W/(m K), respectively, which is higher than the data of 0.074 W/(m K) with the expansion temperature of 800 °C.

2.1.2 ENG-GICs

Owing to economic concerns, expanded graphite is industrially derived from sulphuric acid-based GICs precursor, especially graphite bisulphate. The development of the material involves sulphuric acid with a strong oxidizing agent such as HNO_3 [8], $KMnO_4$, $FeCl_3$ [9], SO_3, O_3, Cl_2 [10], $(NH_4)_2S_2O_8$ [11], etc. GICs with nitric acid and ENG based on them are also widely researched [12]. Apart from graphite bisulphate and nitrate, other precursors also have been applied for ENG preparation. They include GIC with bromine [4], formic acid [13], $FeCl_3$ [14] and ternary co-intercalated compounds K-THF [15], Na-THF, Co-THF [14], HNO_3-organic molecules [16], etc. [17]. Thus, a rich intercalation compound could be obtained, and it needs to be washed thoroughly. The resulting material is poor GIC. By heating this solid, the residual intercalates, and the water molecules inserted during the washing process vaporize, pushing the graphite layers apart: a low-density substance, exfoliated graphite (termed ENG-GICs in this paper), is obtained. The most conventional heating source is the flame, through which the intercalated graphite can be achieved. With such a high temperature, the volatilization of intercalating can also be almost completed together. Alternative methods such as microwave [18, 19], infrared, laser [17], and plasma [20] for ENG-GICs production are described. Also, a chemical reaction between two species in the interlayer galleries of a GIC may lead to exfoliation [21, 22], and the highly porous wormlike accordions of graphite can be achieved.

The structure of such "worms" is described by Celzard in three different scales: the worm itself (Fig. 2.4a) [1], the flattened "balloons" with which the particles are made (see Fig. 2.4b) [1], and the elementary graphite sheets (Fig. 2.4c) [1]. The worm's structure is vital because the worms retain a separate identity in the compressed materials [1, 23]. Indeed, in moderately compacted ENG, such flakes are expected to rearrange and to orientate with each other, and then lead to anisotropic physical property. Conversely, the intermediate scale corresponding to the observed balloons does not deserve so much attention. For a type of given graphite precursor, the sizes, the number, and the distribution of balloons along the worms depend on the intercalate [1]. Figure 2.4b shows different kinds of balloons observed in a given batch of ENG, i.e., for which the worms are prepared from the same starting material with the same intercalate under similar expansion conditions. It shows that some balloons have the same diameter as the worm itself, while some others are divided into many smaller units. Celzard found that the balloons collapse rapidly when the ENG worms are compressed. At the minor scale, a scanning electron microscope presents that the microstructure of ENG is roughly a flattened irregular honeycomb network of graphite (Fig. 2.4c) [1]. The compressed ENG is an axenic material due to such a structure. In other words, a block of ENG becomes thinner in the compression process. A remarkable difference in morphology is observed between the ENG produced by an acceptor- and donor-type GICs. Acceptor-type GICs lead to ENG consisting of giant collapsed balloons with a thickness of 200–300 μm and a width of about 400 μm, while donor-type GICs give many tiny balloons with a thickness of 20–50 μm and a width of 50 μm by SEM data [17].

The properties of the ENG-GICs are related to various techniques for developing the mixture and the exfoliating processes [24]. For example, Malas et al. oxidized

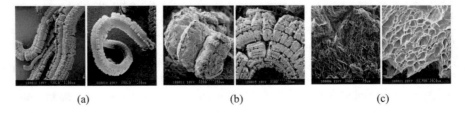

(a)   (b)   (c)

Fig. 2.4 SEM pictures of ENG: **a** Constituting individual worms of raw expanded graphite; **b** "balloons" within the EG worms; **c** distorted honeycomb microstructure [1]

natural graphite by H_2SO_4/HNO_3 (3:1) at 90 °C for 12 h with vigorous stirring, and used ENG-GIC for styrene-butadiene rubber/polybutadiene rubber (SBR/BR) blends. The rubber composites improve thermal stability and hardness [25]. Yue et al. mixed 10 ml of concentrated sulphuric acid (98%) and 1 ml of hydrogen peroxide (30%) with 6 g of natural flake graphite (35 mesh, 99% purity, 97.5% crystallinity) at room temperature. The mixture is placed for 90 min, washed with water to pH 5–7, and dried at 80 °C for 24 h to form expandable graphite. The expandable graphite as prepared is heated at 600 and 1000 °C for 15–20 s, forming ENG-GIC-1 and ENG-GIC-2, respectively. They investigated the difference between the microstructure evolutions of the two ENG-GICs by high-energy ball-milling, and the results presented that the flexibility of the thin graphite sheets of ENG-GIC-2 is higher than that of ENG-GIC-1 [26].

Han et al. studied the expanding technique for the expandable graphite that is treated by sulfuric acid [27]. The significant elements in the expandable graphite are C, S, and O. The amount of the S is 2.8 wt% because the sulfuric acid has been used for treating the sample. They expanded the expandable graphite at different temperatures and found the relation between the residual sulfur and the weight percentage of carbon after the thermal treatment. Since the water, sulfuric acid, and other volatile compounds are removed, relative carbon contents increase with increasing heat treatment temperature, as shown in Fig. 2.5. They suggest that a graphite sulfate is formed with the chemical formula of $C_8^+(OH)_3H_SO_4\text{-}2H_2SO_4$. The sulfuric acid can be easily decomposed to yield the sulfuric acid and original graphite in the free state as the bonds are ionic. Since a trace amount of sulfur can cause the corrosion between the solid sorbent and the metallic walls, the less of sulfur in the ENG-GICs is the better for the solid sorption system.

The ENG manufactured by the Mersen in France is researched extensively, and the sulphuric acid is adopted for forming GIC. The technical expansion process is as follows: Firstly, the natural graphite platelets are treated with sulfuric acid, and

Fig. 2.5 Change of carbon and residual sulfur contents with heat treatment temperature [27]

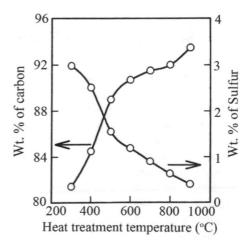

then the resulting intercalation compound is submitted to a thermal shock by passing through a flame. The sudden volatilization of the intercalate makes the platelets swell drastically, and as a result, the very light distorted accordion-like particles can be obtained. Such a commercialized product is defined as ENG-TSA to distinguish it from ENG and other types of ENG-GICs that are developed in the lab. Raw ENG-TSA has very low apparent density [28]. The property of such material is shown in Table 2.1. The carbon content in the material is higher than 99.8%, and the ash content is less than 0.2%. The density of the material is only 5–6 kg/m^3, which is about three times less than that of standard ENG [29].

2.2 Techniques for Developing the Composite Solid Sorbents with ENG

The techniques for developing the composite solid sorbents with ENG as matrix are mainly as follows:

(1) Simple mixture and consolidation: sorbent and the additive are mixed by a defined mass ratio or volume ratio [30] and then are consolidated by the compression process. This method is mainly used for the sorbents that cannot be dissolved in the water, such as the AC. Sometimes it also has been used for silica gel.

(2) Impregnation and compression: procedures include dissolving the sorbent in the water, impregnating the additive with the solution, then compressing the mixture, and lastly, drying the composite materials. One advantage of such a type of sorbent, especially the silica gel, is the large porosity, which benefits the mass transfer performance [30]. The impregnation process also could guarantee the uniform distribution of the solid sorbents in the matrix.

2.2.1 Simple Mixture and Consolidation

2.2.1.1 Physical Composite Solid Sorbents

For producing the consolidated composite sorbent of AC, firstly, the AC and ENG are simply mixed, and then the mixture is compressed and consolidated [31, 32]. The smaller the ratio between AC and ENG is, the easier it is to be consolidated. Generally, there are no cracks on the sorbent when the ratio of AC is 1:2, as shown in Fig. 2.6a. Cracks begin to occur on the surface of the consolidated sorbent when the mass ratio between AC and ENG increases to 1:1 (Fig. 2.6b). The number of cracks increases when the ratio of AC is larger, as shown in Fig. 2.6c. The largest ratio between AC and ENG is 2:1, and the sorbent with the ratio larger than 2:1 could not be consolidated (Fig. 2.6d).

Table 2.1 The composition of ENG-TSA [29]

Carbon content (%)	Ash content (%)								Resistance to temperature in an oxidizing atmosphere	Density (kg/m^3)
>99.8	<0.2								500 °C	5–6
	Average ash content (ppm)									
	Fe	Si	Mg	Al	Ca	Cr	Mn	Cu		
	150	230	60	20	10	10	10	10		

(a)   (b)   (c)   (d)

Fig. 2.6 The solidified composite sorbents with different ratios between AC and ENG: **a** 1:2; **b** 1:1; **c** 2:1; **d** 2.5:1 [31]

Jin et al. described the procedures for developing the composite compact AC in detail [33]: Firstly, the natural graphite is expanded under the temperature of 600 °C for about 10 min, then a certain amount of water is mixed into the quantified ENG and AC together, and at last, the ENG-water-AC mixture is dried in the oven at 120–130 °C for 12 h. In experiments the consolidated AC with ENG at the density of 600 kg/m^3 shows the best heat transfer performance, and their thermal conductivity varies from 2.08 to 2.61 W/(m K) according to the fraction of AC.

Wang et al. developed the composite AC with the matrix of ENG-TSA by simple mixture and consolidation, i.e., consists of mixing AC, water, and ENG-TSA, and mechanical compression of the composite sorbent [34, 35], which is shown in Fig. 2.7. Firstly, the granular ENG-TSA and granular AC are all dried in the oven at the temperature of 150 °C for 2 h to eliminate the mass error caused by moisture variation (Fig. 2.7a, b). After that, the AC is mixed with water by the mass ratio of 1:1 to make AC moist, which makes the mixing process easier with the granular ENG-TSA (Fig. 2.7c, d). The composite material is then compressed by a pressure controlled between 0–10 MPa (Fig. 2.7e), and finally, the consolidated samples are dried in the oven at the temperature of 150 °C for 4–5 h (Fig. 2.7f). The sample ready for the test is shown in Fig. 2.7g.

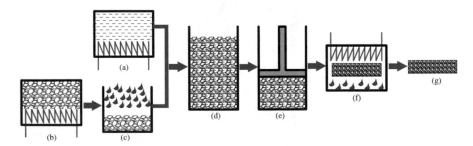

Fig. 2.7 Development of consolidated composite AC: **a** drying process of granular ENG-TSA; **b** drying process of granular AC; **c** mixture of granular AC and water; **d** composite sorbent of AC, ENG-TSA, and water; **e** compressing process of composite sorbent; **f** drying process of the consolidated sorbent; **g** consolidated composite sorbent for the test [34]

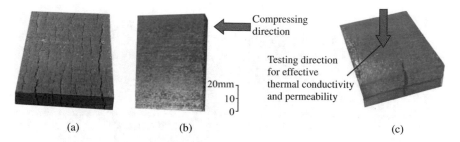

Compressing direction

Testing direction for effective thermal conductivity and permeability

20mm
10
0

(a)   (b)   (c)

Fig. 2.8 Consolidated composite sorbents of AC: **a** density of 249 kg/m^3 with cracks; **b** density of 388 kg/m^3 without cracks; **c** density of 448 kg/m^3 with cracks [34]

The consolidating process for the composite AC/ENG-TSA is complicated because AC is hard and cannot be dissolved into the water. Samples with high AC content, too big or too small density, or large AC grains, are fragile and likely to crack. For example, the consolidated samples are shown in Fig. 2.8, in which the size of AC is 80–100 mesh, and the percentage of AC is 67%. The sample with a density of 338 kg/m^3 has no cracks (Fig. 2.8b). Whereas the cracks occur for the density of 249 kg/m^3 (Fig. 2.8a) and the density of about 448 kg/m^3 (Fig. 2.8c).

A similar method is also utilized for the composite sorbent of silica gel and ENG-TSA. The procedures are as follows [36]: Firstly, both the granular ENG-TSA and silica gel are dried in the oven at the temperature of 120 °C for 4 h. Silica gel is then moisturized with water by the mass ratio of 1:1 before mixing with ENG-TSA for uniformity in the mixture. The composite of silica gel, ENG-TSA, and water is then compressed. Finally, the consolidated samples are dried in the oven at the temperature of 120 °C for 4–5 h.

The composite sorbents of silica gel/ENG-TSA are also hard to be compressed when the content of silica gel is too large and the density is small. Therefore, blocks with silica gel percentage higher than 80% could hardly be produced due to the sharp decline in the bulk density of ENG-TSA. For instance, Fig. 2.9a, b show two samples with similar bulk density but a different proportion of silica gel. No crack happens for the sample with silica gel content of 67%, whereas the sample with the silica gel content of 80% fragments.

Fig. 2.9 Consolidated composite sorbent of silica gel: **a** silica gel content of 80%; **b** silica gel content of 67% [36]

(a)   (b)

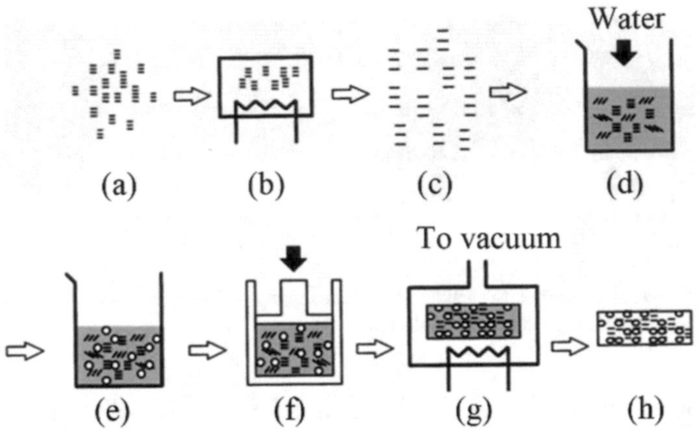

Fig. 2.10 Manufacture procedure of composite blocks: **a** expandable graphite powder; **b** heat treatment for expansion; **c** expanded graphite powder; **d** slurry of expanded graphite and water; **e** mixture of the slurry and silica gel; **f** pressing; **g** vacuum drying; **h** composite block [37]

The manufacturing procedures of the composite sorbents developed by Eun et al. [37] are presented in Fig. 2.10. They used sulfuric acid as an intercalating material. Sulfuric acid is thermally decomposed at 340 °C due to hydrating of sulfuric acids, such as monohydrate and dihydrate, with boiling points below 300 °C [37]. Higher temperature leads to fast volatilization and the decrease in sulfur content, and it is below 1% for the higher temperature above 600 °C. Thus, the ENG-GIC manufactured at 600 °C for 10 min is used for composite blocks as it has short heating time, low sulfur content, and a small amount of carbon loss by combustion. The expanding process of ENG-GIC is shown in Fig. 2.10a, b, c. The silica gel powders are added to the slurry of ENG-GIC and water by the ratio of 5 ml water/1 g graphite (Fig. 2.10d, e). The well-mixed mixture is compressed (Fig. 2.10f) and then is dried at 80 °C for 1 h. Finally, the composite block is developed after removing water entirely in a vacuum at 145 °C for more than 2 h (Fig. 2.10g, h).

Both composite AC and silica gel developed by the simple mixture and compression have been well embedded in the matrix, as shown in Fig. 2.11a–d. The SEM pictures also indicate the organized structure of ENG-TSA at higher density

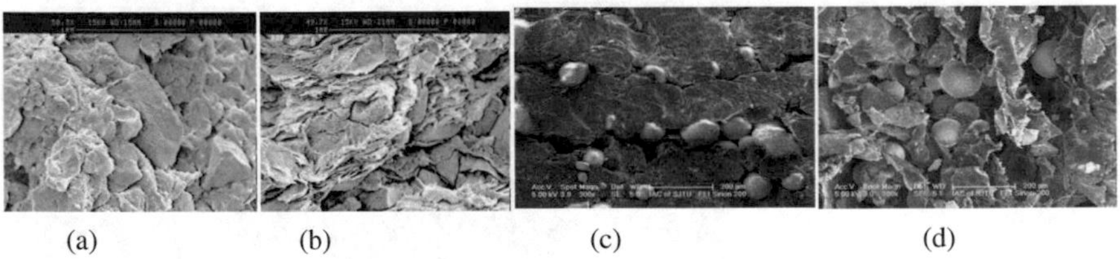

Fig. 2.11 SEM pictures of consolidated composite sorbents, **a** AC 33%, 278 kg/m^3; **b** AC 67%, 448 kg/m^3, **c** silica gel 50%, 269 kg/m^3, and **d** silica gel 67%, 378 kg/m^3 [34, 36]

(Fig. 2.11b). Such structure could improve the heat transfer performances and will lead to the anisotropic heat and mass transfer performance.

2.2.1.2 Chemical Composite Sorbents

Lee et al. developed the non-uniform reaction blocks for the chemical heat pump [38]. Firstly, the expandable graphite treated with acid is expanded at 800 °C for 10 min. Optimal expanding conditions are set by mass variation, bulk density, and pH of graphite. After the expanding stage, the ENG-GIC is cleaned by distilled water to remove remaining contaminants. Then a rotary vacuum evaporator is employed to remove water and air in the pore of ENG-GIC at 65 °C for 1 h. Then, halides of $NiCl_2$ and $MnCl_2$ are well mixed for 6 h. A compressive molder is specially designed for producing the dense reaction blocks with ENG-GIC powders, which is shown in Fig. 2.12. The non-uniform reaction block is manufactured by the modeler in Fig. 2.12 to increase the apparent density with the radius of reaction blocks. The ENG-GIC mixed with halides is filled inside the modeler to the height from h_1 to h_5 in Fig. 2.12, and then the sample is compressed by the external force. The density of the central cylinder is fixed to 165 kg/m^3, and then the data changes, and it is 222, 279, 337, and 394 kg/m^3 gradually from the inside to the outside. Lee et al. indicated that the non-uniform reaction blocks have much better heat transfer capability since the temperature gap between the inner and outer is smaller than that of uniform reaction blocks, and such a phenomenon is helpful for the improvement of the reaction rate and the decrement of the cycle time.

Fig. 2.12 Molder for non-uniform reaction blocks [38]

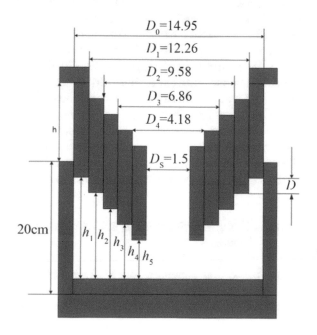

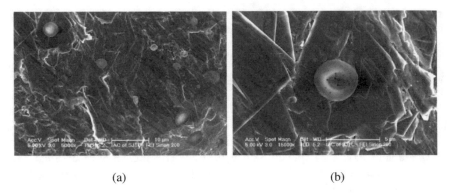

(a)   (b)

Fig. 2.13 SEM pictures of consolidated composite sorbent: **a** 5000 magnification; **b** 15,000 magnification [40]

Kim et al. studied the mixture of ENG and magnesium hydroxide ($Mg(OH)_2$) to enhance the thermal conductivity and reactivity of magnesium oxide/water (MgO/H_2O) chemical heat pump [39]. They mix the $Mg(OH)_2$, $CaCl_2$, and ENG along with 100 ml of ethanol in the flask of a rotary evaporator, and then let the ethanol evaporate at 30 °C. At last, the mixed materials are dried in a drying oven at 120 °C for over 10 h. Experiments show that the reaction rate for the dehydration of materials is improved dramatically by the enhancement of the heat transfer and mass transfer. Simultaneously, the mean heat output rate by hydration is also increased, and the performance is better than that of chemical heat pumps with other materials.

Jiang et al. studied composite sorbents with the matrix of ENG-TSA [40]. Firstly, the ENG-TSA is dried in the oven at a temperature of 120 °C. Secondly, the $CaCl_2$, water, and ENG-TSA are mixed. The composite is dried at the temperature of 120 °C for 4 h to make sure no retained water there and then it is dried at the temperature of 260 °C for 4 h to remove the crystal water. Finally, the composite is compressed by a pressing machine into a block. The SEM picture of the composite sorbent is shown in Fig. 2.13. Figure 2.13a, b demonstrate that the $CaCl_2$ is embedded in the host matrix of ENG-TSA tightly. Figure 2.13 also shows that the compressed material has the structure of micro sheets, which regularly distribute in the consolidated sorbent. Such a structure will enhance the heat transfer performance along the direction of micro sheets and improve the mass transfer performance because the gas will be easier to be transferred in the microchannels through micro sheets.

2.2.1.3 Composite Sorbents of ENG/Carbon Fiber (CF)/Chemical Sorbents

Zajaczkowski et al. combined both ENG and CF and molded them together to obtain the desired shape and apparent density [41]. The ENG ensures heat transfer and allows the forming of new composite into the desired shape. A single ring-shaped

Fig. 2.14 Molding of the ENG/CF composite [41]

sorption module has a diameter of 10 cm and the height between 0.7 and 1.5 cm (Fig. 2.14a, b). The diameter of the centrally located gas diffuser is 1 cm (Fig. 2.14c). Development of materials is conducted under carefully predefined conditions to ensure repeatability and uniform characteristics of different modules, and mass of metallic salt, the parameters of the mass ratio between ENG and fibers, compression force and rotational speed of mixer are controlled. Different compositions of the materials are prepared and tested (CF/ENG ratio 0.25–1.0, inert materials to salt ratio 0.25–0.5, molding force 500–900 kN). Experiments reveal the significant mechanical limitation of CF. Too much pressure makes fibers lose their structure and turn into carbon dust, which makes it impossible to be formed into a usable solid shape. Proper determination of the CF/ENG ratio should be studied for allowing higher molding forces.

2.2.2 Impregnation and Compression

The impregnation and compression are always used for the chemical sorbents that can be dissolved in the water.

Mauran's research group [42, 43] firstly used such a method for the development of the composite sorbent of $CaCl_2$ and ENG-GIC for the improvement of the thermal conductivity, i.e., IMPEX. The manufacturing processes of the IMPEX are shown in Fig. 2.15. Firstly, the ENG-GIC is prepared (Fig. 2.15a–c), and the ENG-GIC block of $\varphi 20 \times 10$ mm is produced (Fig. 2.15d). Then the $CaCl_2$ solution is prepared with $CaCl_2$ percentage of 20% and water of 80%. The graphite block is put into the salt solution, and the container filled with sample and solution is evacuated to vacuum, which can ensure salt solution as much as possible immersed into the graphite block (Fig. 2.15e). After that, the block is immersed in the salt solution for 5 h at the temperature of 100 °C. Lastly, vacuum conditions of 200 °C and 0.02 bar are kept for 3 h to dry the sample (Fig. 2.15f). The formed IMPEX is shown in Fig. 2.15g.

SEM of IMPEX is shown in Fig. 2.16. The needle-like structure is calcium crystal. $CaCl_2$ is distributed uniformly in the block because of the combination of the vacuum condition and impregnation.

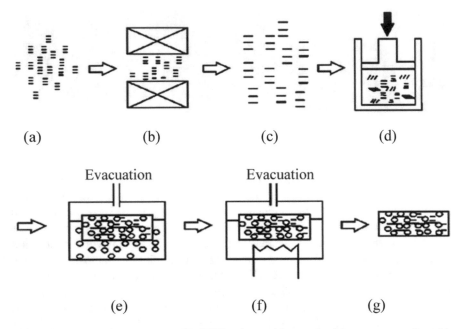

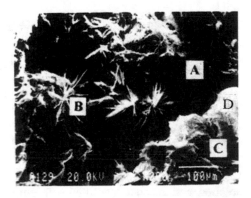

Fig. 2.15 The manufacture process of IMPEX: **a** the graphite powder; **b** heat treatment of graphite powder for expansion; **c** expanded graphite powder; **d** pressing process; **e** immersed process of graphite into $CaCl_2$ salt solution; **f** evacuate to vacuum and dry process; **g** formed IMPEX [42]

Fig. 2.16 The salt distribution in IMPEX block, A-space inside the graphite; B-Ca crystals; C-the graphite; D-the distribution of Cl [42]

When reacted with the ammonia, the volume expansion rate of IMPEX is 0 when the volume density of the graphite is 156 kg/m^3, and such a phenomena means that the size of IMPEX does not change though the expansion and agglomeration of $CaCl_2$ occur during the sorption/desorption processes due to the porous ENG in the mixture.

The procedures studied by Mauran are complex because the vacuum is required, compared with that the developing procedures proposed by Wang et al. are quite simple. The main difference lies in the manufacture of the sorbent block. Mauraun group has manufactured the graphite block first and then immersed the graphite into the salt solution. Wang and Olivera immersed the graphite into the salt solution first and then compressed the composite into the block [44, 45]. They firstly expand the 80 mesh expandable graphite at 800 °C for 2 min. After that, the ENG is immersed into the salt solution with $CaCl_2$ mass of 14%, and then the samples are dried at 110 °C for 22 h and dried at 270 °C for 8 h to remove the water and ensure $CaCl_2 \cdot nH_2O$ stabilized. Lastly, the composite is compressed into the block under the pressure of 10 MPa in the mold. The SEM picture of the sample is shown in Fig. 2.17. The graphite block appears the parallel layer of flake-like in the radius direction, which is perpendicular to the compression direction, and this direction is also the optimal direction for heat and mass transfer. Such a method also has been used for the LiCl/ENG [46], $MnCl_2$/ENG, and NH_4Cl/ENG sorbents [47].

Fujioka et al. studied the composite sorbent of $CaCl_2$ and ENG [48]. They firstly soaked and rinsed ENG in ethanol, and then rinsed it with water and soaked it in $CaCl_2$ aqueous solution. After that, they heated it at around 150 °C to remove excess water until the moles of H_2O in 1 mol of $CaCl_2$ was about 3. The mass ratio of $CaCl_2$ to expanded graphite is 8:1. SEM pictures of ENG and composite material are presented in Fig. 2.18. Porous ENG of several microns is separated by thin

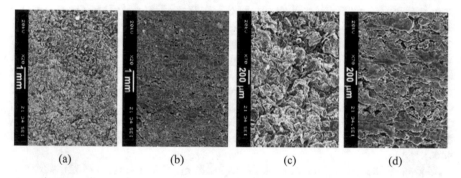

(a) (b) (c) (d)

Fig. 2.17 The SEM figure of solidified sorbent: **a, c** parallel to the compression direction; **b, d** perpendicular to the compression direction [45]

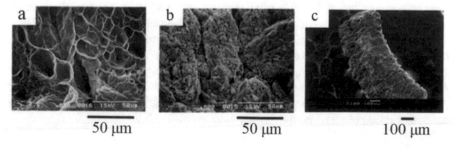

50 μm   50 μm   100 μm

Fig. 2.18 SEM photographs: **a** ENG; **b** inside of the composite reactant; **c** whole shape of a composite particle [48]

graphite walls (Fig. 2.18a), and it is suited for preparing composite materials with large surface area, high thermal conductivity, and permeability. Figure 2.18b is an SEM photograph for the inside of composite sorbent, which shows that fine grains of $CaCl_2$ has been deposited in the pores of ENG. The shape of a particle of composite reactant likes a worm about 1 mm in length, as presented in Fig. 2.18c.

Kim et al. [49] use the ethanol as the media for the impregnation processes, and the composite sorbents of Mg(OH) and $CaCl_2$ reacted with the refrigerant of water are studied. They dissolve $Mg(OH)_2$ and $CaCl_2$ into a flask with 100 ml ethanol and then let the mixture be dispersed in ethanol using an ultrasonic processor. After that, the measured ENG is charged in a flask. The flask is set in a rotary evaporator and rotated at 30 °C under pressure lower than the atmosphere. Ethanol is removed from the flask. At last, the mixed material in paste state in the flask is dried in an oven at 120 °C for over 10 h. The final product is consolidated composite sorbent, which has better performance on dehydration than pure $Mg(OH)_2$.

The impregnation method developed by Py et al. is quite complex [50, 51]. They thought that the impregnation has advantages over the physical mixing method because of the better distribution of chemical agents into the carbonaceous particle mass. They utilized $ZnCl_2$ as the activation for almond shells by impregnating in solution. Firstly, they soaked the almond shell powders in stirred (200 rpm) aqueous solutions of various $ZnCl_2$ contents (up to 0.3 kg/l) at 85 °C over 21 h. The slurry was filtered, and the solid phase was kept at 120 °C over 12 h under air atmosphere. The impregnated precursor was mixed to vermicular ENG, and the mixture was molded and compressed at 80 bar. Then the composite was activated with the nitrogen in the horizontal furnace from room temperature to 600 °C at 4 °C/min and a 600 °C plateau over 5 h. The obtained composite was rinsed with HCl solution at 85 °C over 24 h, with distilled water at 80 °C and dried at 120 °C. Such situ activation of different precursors within a consolidated ENG matrix led to composites of high thermal conductivity (from 1 to 32 W/(m K)) and high effective sorbent level (80 wt%).

Gao et al. developed the compact composite multi-salt sorbents [52], which is proved for decreasing the hysteresis phenomena. Firstly, different metal chlorides and matrix, such as NH_4Cl, $CaCl_2$, $MnCl_2$, and ENG-TSA, were dried in the oven for 3–4 h to dry up the water inside the sorbents and matrix. The temperature of the oven was controlled at 130 °C. After that the salts were weighted by the balance with accuracy of 0.01 g, and then they were dissolved in water, respectively. Then, the ENG-TSA was mixed in the solution. For the bi-salt composite of NH_4Cl and $CaCl_2$, the ratio among NH_4Cl, $CaCl_2$, and ENG-TSA were 2:2:1, and for the tri-salt mixture, the ratio among NH_4Cl, $CaCl_2$, $MnCl_2$, and ENG-TSA was 4:4:4:3. After the composite was uniformly mixed, the mixture was placed in an oven and dried 6–7 h at the temperature of 160 °C. Finally, the dried composite sorbent was compressed into blocks, and the density is about 400 kg/m^3.

2.3 Techniques for Developing the Composite Solid Sorbents with Activated Carbon Fiber or Activated Carbon

2.3.1 The Characteristics of Active Carbon Fiber

Activated carbon fiber (ACF), which is made of the fibrous precursor body via the carbonization and activation, is also called fibrous activated carbon. Compared to the activated carbon, it has better sorption performance. Meanwhile, it is an environmentally benign engineering material and has excellent thermal conductivity [53]. More than 50% of the carbon atoms are situated inside and outside the surface, forming a unique adsorptive structure. Moreover, it has a large specific surface area and narrow pore size distribution, which makes it possess the higher sorption/desorption speed and larger sorption capacity.

(1) Structure characteristics

Activated carbon fiber is a kind of typical microporous activated carbon (MPAC). It is considered to be the combination of "ultra micron particle, irregular surface structure, and very narrow space" with a diameter of 10–30 μm. The pore is located on the fiber surface. The ultra micron particle is combined with different ways, forming the productive nano space. The nanometer space scales have the same order of magnitude with the size of the ultra micron particle and lead to the large specific surface area. It contains many irregular structures, such as the heterocyclic structure or the microstructure of the surface functional groups. The interaction between the micropore and molecular in the pore side will create a robust molecular field because of the excellent surface energy, and thereby providing molecules under the sorption state with the high-pressure system of the physical and chemical change. As a result, the diffusion path to the vacancy of the sorbates is shorter than that of activated carbon, and the driving force is larger, and pore size distribution is very concentrated. It is the main reason that ACF has the larger specific surface area, the faster desorption rate, and the higher sorption efficiency than activated carbon.

(2) Functional methods

Using pore interstitial structure control and surface chemical modification can achieve the efficient sorption transformation of the specific materials. ACF is usually suitable for the sorption of gas and liquid molecules with low relative molecular mass (relative molecular mass < 300). When sorbent pore size is two times as much as the critical size of the sorbate molecular, the sorbate is easily sorbed. Aperture adjustment can make the micropores of ACF match the molecular size of the sorbent. There are usually three methods: (a) Using the activation process or changing the activate degree to the nanometer level to adjust aperture; (b) Adding metal compounds either in the raw fiber or ACF and activation can adjust the pore size. Besides, the carbonization and activation of some other materials can change the pore size. The raw fibers need to have relatively big pore size; (c) Using the hydrocarbons and deposition in the cell wall, post-processing under high-temperature conditions can decrease pore size. Surface chemical modification can vary the acid and alkaline of the ACF surface by introducing or removing some surface functional groups. After the high temperature or the hydrogenation, the surface oxygen groups can be removed. Through the gas-phase oxidation and liquid-phase oxidation, the acid surface can be obtained. The influence of the physical and chemical structure should be considered during the modification process.

In the sorption refrigeration applications, composite sorbents of activated carbon fiber and chemical sorbents have many advantages. On the one hand, the excellent thermal conductivity can improve the heat transfer performance of chemical sorbents; on the other hand, the rich microporous structure can suppress the expansion and agglomerate phenomenon of chemical sorbent to enhance the mass transfer performance.

2.3.2 Composite Sorbent with Activated Carbon Fiber as Matrix

The activated carbon fiber (ACF) chosen as a matrix of composite sorbent will improve the heat transfer of chemical sorbent effectively. Generally, the higher graphitization degree of the activated carbon fiber will lead to better heat transfer performance. However, since the price of activated carbon fiber is quite expensive, the optimal activated carbon fiber always should have a comparatively low price as well as a high graphitization degree.

Currently, two types of successful composite sorbents with activated carbon fiber as an additive are the composites of activated carbon fiber and $MnCl_2$ (ICF) [54, 55], as well as activated carbon fiber and $CaCl_2$(GFIC) [56–58]. The first one is called ICF (Impregnated carbon fibers with $MnCl_2$) [54, 55]. The procedures for preparing the composite sorbents of activated carbon fiber and $MnCl_2$ by impregnation method are as follows: Firstly, dilute the $MnCl_2$ with alcohol, and then impregnate the activated carbon into the solution of $MnCl_2$, after that heat the composite sorbent for drying the sample. For such type sorbent (ICF), the combination of $MnCl_2$ and activated carbon fibers is more firmly after the sorbent is prepared. The preparation time of ICF is short, which takes only 3–4 h. However, because ICF and $MnCl_2$ only remain at the micromolecular level, after several cycles of heating desorption and cooling sorption, the combination between halides in the ICF and activated carbon fiber will break. Then halide will fall from the activated carbon fiber and will accumulate at the bottom of the reactor, which will influence the sorption and desorption performance. The second one is called GFIC (Graphite fibers intercalation compounds) [54, 55]. Firstly, activated carbon fiber is graphitized at high temperatures (thermal conductivity after graphitization is higher than 600 W/(m K)), and then under the conditions of chlorine gas temperature of 500 °C $MnCl_2$ is intercalated in carbon fibers. To gain a high

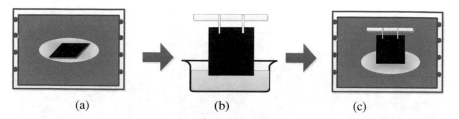

Fig. 2.19 Developing procedures of the composite sorbents with ACF as a matrix. **a** The ACF is dried in the oven, **b** impregnation **c** The ACF is dried in the oven after impregnation [59]

intercalated rate, in the process of preparing composite sorbent, a small number of other types of halide will be added. The preparation time of the GFIC is long, generally taking about one week. GFIC in $MnCl_2$ particles of activated carbon fiber keeps at the atomic level. Even if some of the halides are inclined to leave the activated carbon fiber sandwich, the combination of halide and activated carbon fiber is very firmly.

In recent years Wang et al. developed the ACF with a thickness of 1 mm used for the $CaCl_2$ composite sorbents [59]. First, the ACF felt was cut into 6 cm × 6 cm pieces, then dried in an oven with a temperature of 120 °C for 4 h until the mass does not change. Then the ACF felts were sealed and cooled to room temperature. The four pieces of ACF felts were impregnated in the aqueous solution with a mass concentration of 10, 20, 30, 40% $CaCl_2$, respectively, with a temperature of 20 °C for 8 h. The picture is shown in Fig. 2.19. Because of the extremely capillary force of ACF felts, the liquid can be sorbed from the bottom to the top of the samples, and the sufficient impregnation length is 6 cm. Due to fiber weaving, the ACF material is anisotropic. To study the impact of the direction of the fibers on impregnation, two pieces of ACF felt impregnating in 30 wt% solution were developed, one with the fibers direction vertical to the solution level (ACFV), and another with the fibers direction parallel to the solution lever (ACFP). Three groups were set in order to reduce experimental error.

The material of the ACF felt is soft and flexible and looks like a piece of thick cloth. It turns to become stiffen after impregnation of the salt and drying process. Figure 2.20 shows the appearance of ACFV and ACFP. The white patterns of ACFP and ACFV, on behalf of $CaCl_2$, are different from each other. Figure 2.20 also shows that halide decreases with increasing height.

To eliminate the individual differences, three groups were set to research the effects of the fiber direction on salt impregnation. It is shown in Fig. 2.21 that the halide increments of three specimens are close (in the vicinity of 0.86), and the ratio of ACFP30 is a little higher than ACFV30.

A SEM was adopted to observe the microscopic morphology of samples. For ACF felts, it can be observed in Fig. 2.22a that a large number of fibrils, which are in irregular shape and different sizes, array axially, create a small amount of discontinuous texture groove and wedge-shaped axial crack at the macro level. The

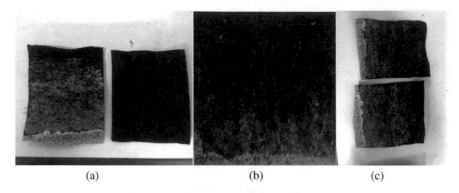

(a)   (b)   (c)

Fig. 2.20 Appearance of ACF before and after impregnation **a** ACF, **b** ACFV, **c** ACFP [59]

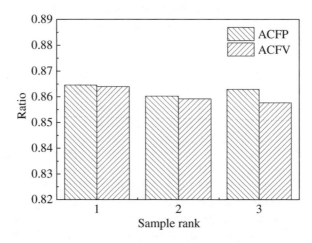

Fig. 2.21 Effect of fiber orientation on impregnation [59]

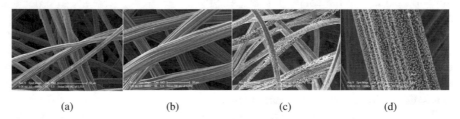

(a)   (b)   (c)   (d)

Fig. 2.22 SEM pictures of ACF felt before and after impregnation; **a** 1500 times magnification of ACF, **b** 3000 times magnification of ACF, **c** 1500 times magnification of ACF-CaCl$_2$, **d** 12,000 times magnification of ACF-CaCl$_2$ [59]

microstructure of ACF felts increase the surface area, enlarge the capillary force of the materials, and ensure being an excellent carrier to conduct and hold water.

After impregnating in the concentration of 30% calcium chloride aqueous solution (ACFV30), shown in Fig. 2.22c, d, fiber surface uniformly covered with calcium chloride as the solid sorbent for sorbing water vapor in the air, and the space formed by the fibers can accommodate more liquid. The solid calcium chloride sorbs water into a concentrated solution, which also has the strong sorption capacity of water. The calcium chloride attached to the matrix is composed of a large number of the fibril, and the tremendous capillary force of the ACF felts ensures that plenty of fluids will not flow out.

2.3.3 Composite Sorbent with Activated Carbon as Matrix

Wang et al. apply activated carbon as a matrix to halide-ammonia working pairs, aiming at improving the mass transfer of chemical sorbent by adding activated carbon. Heat transfer will be further improved if the composite is consolidated [60].

Compared with sorbent with graphite and activated carbon fiber as a matrix, sorbent with activated carbon as a matrix takes the advantages of the hard activated carbon particles as well as the simple preparation process of composite sorbents. Because of hard activated carbon particles, the chemical sorbent in activated carbon will not detach from the porous media and accumulates at the bottom of the sorption bed, which generally occurs for the matrix of expanded natural graphite and activated carbon fiber. Such a process means that the chemical sorbent will be evenly distributed in the activated carbon particles.

For different application occasions, composite sorbents of activated carbon and $CaCl_2$ can be chosen as a bulk mixture as well as consolidated composite. The bulk mixed sorbent is generally used for sorption bed with limited filling space such as the plate sorption bed because consolidated mixed sorbent is hard to fill. The consolidated composite sorbent can be used for the sorption bed with ample filling space such as a tube sorption bed.

For the preparation of bulk sorbent, there is little difference between the composites made by impregnation and the simple mixing process. Generally, the simple mixing procedure is used for the preparation of bulk sorbent. The process is to mix the activated carbon and $CaCl_2$, and then dried it for the later application. For preparing the consolidated sorbent, $CaCl_2$ is dissolved into the water, and then cement and activated carbon are added in the solution, and lastly, the composite will be consolidated and dried. The simply mixed composite sorbent and consolidated sorbent prepared by mold are shown in Fig. 2.23 [61].

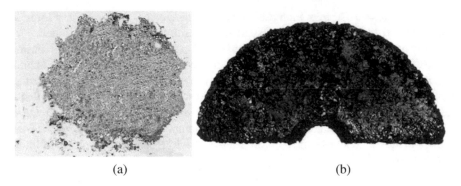

(a)   (b)

Fig. 2.23 Composite sorbent activated carbon and CaCl$_2$ [61]. **a** Simply mixed composite sorbent; **b** Consolidated composite sorbent

2.4 Techniques for Developing the Composite Solid Sorbents with Silica Gel

Composite sorbents with silica gel as matrix are generally prepared by the impregnation method [62–65], for which silica gel is soaked in salt solutions (such as CaCl$_2$) with a fixed solubility, and then dry the silica gel to get composite sorbent with strong sorption capacity. Composite sorbent with silica gel as a matrix is to impregnate the halide chemical sorbent into the microspores of the silica gel to get the strong sorption capacity. In actual use, halides solution after sorbing water is required not to flow out of pores of silica gel, for which the unique processing technique is needed. The characteristics of the composite sorbent can be improved by the following procedures [66]:

(1) Change the porous structure of the matrix. In silica gel-halide composite sorbent, each atom is generally closely surrounded by the atoms with characteristic diameter. If the diameter of halide atoms in the solution is far less than the atom diameter of silica gel, the micropores of silica gel will not affect the distribution of the halide, and the sorption performance trend of a composite sorbent is similar to the inorganic salts. Otherwise, its sorption performance will be different from that of inorganic salt.

(2) Change the chemical characteristic of intercalation halides. For various inorganic halides, the complexion reaction between the halides and water will be different. The performance of composite sorbents will show different trends.

(3) Change the number of halides inside micropores. Generally, agglomeration and swelling phenomenon of sorbent will become serious when the mass of halide in composite sorbent increases. This will improve the heat transfer performance as well as influence mass transfer performance.

The preparing processes of silica gel/CaCl$_2$ composite sorbent in Shanghai Jiao Tong University are as follows [67]:

(1) Preparation for $CaCl_2$ aqueous solution. Dissolve the $CaCl_2$ into the water with the designed concentration of the halide. If the required concentration of the solution is not very high, $CaCl_2$ can almost instantly dissolve. For calcium chloride aqueous solution with high concentration, it commonly takes about 30 min to prepare.

(2) Sample preparation. Soak the silica gel as the matrix in $CaCl_2$ aqueous solution, and then put the sample in the environment for 12 h. Such a process makes $CaCl_2$ impregnate into the micropores. Then filter the silica particles filled by $CaCl_2$ with the sieve. The rest of the low concentration of $CaCl_2$ solution will not be used. Heat and dry the silica gel particles at 80 °C in constant temperature and humidity oven. In the drying process, weigh the samples at regular intervals, and stop the drying process when the weight of the sample no longer reduces.

Aristov produced composite sorbent for testing thermal conductivity, for which the preparation method is as follows [66]: Mix silica gel powder and $CaCl_2$ solution of 40% concentration. Solidify the powder into a mold with the shape of square brick with a size of $7 \times 3 \times 1.5$ cm^3. Contact the brick with the water vapor till a pre-determined water sorption value is gotten and then test its thermal conductivity.

According to the testing results, the highest sorption capacity of silica gel-$CaCl_2$ composite sorbent with water can reach 0.7–1.5 kg/kg, which is further more than that of zeolite to water as well as silica gel to water.

However, there is always an adverse effect on the preparation process of silica gel-halide composite sorbent, i.e., the destruction of the solid skeleton of the silica gel particles by the impregnation of the $CaCl_2$. This is because sorption heat leads to the rapid rise of the temperature of the silica gel. Local temperature can even reach more than 120 °C, which will destroy the structure of the silica gel. Figure 2.24 shows the inner structure of the composite sorbent after sorption and desorption via an electron microscope. Results show that the higher concentration of $CaCl_2$ is the greater degree of fragmentation is. Thus, pure silica gel will show better sorption performance than the composite sorbents when relative humidity is low. If the sorption performance

Fig. 2.24 SEM (scanning electron microscope) picture for composite sorbent [67]

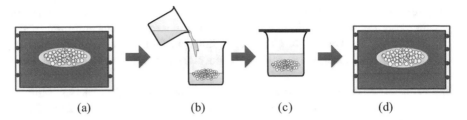

(a)   (b)   (c)   (d)

Fig. 2.25 Developing procedures of the composite sorbents with silica gel as a matrix. **a** The SC is dried in the oven, **b** impregnation **c** standing for 48 h **d** the SC is dried in the oven after impregnation [59]

of the composite sorbent is required for a better performance than that of silica gel, the concentration of CaCl$_2$ aqueous solution should be higher, which can improve the sorption performance of silica gel by the additive of salts. What is more, relative environmental humidity should be lower than 70% for composite sorbent. Otherwise, the liquefaction of CaCl$_2$ will influence the performance of composite sorbent.

Wang et al. studied silica gel, named type C (SC) with the average pore diameter of 8–10 nm, and it was selected as the water sorbent matrix for air-to-water sorption processes [59]. First, silica gel was dried in the oven at the temperature of 120 °C for 4 h until the mass does not change. Then silica gel was sealed and cooled to room temperature. After that, the silica gel was immersed into calcium chloride aqueous solution with a different mass concentration of 9, 16, 23, and 30% at room temperature for 48 h (Fig. 2.25). A vacuum suction filter bottle connected with a vacuum pump was used to separate the liquid and samples, to remove the solution on the gap and surface of silica particles. At last, the composite samples were dried again in the oven at 120 °C until no weight reduction.

With the increase of the concentration of the calcium chloride aqueous solution (Fig. 2.26), the color of the dry sample turns from semi-transparent to white, which indicates the more amount of calcium chloride impregnating in silica gel.

What is more, the microstructure of composite samples with type C silica gel as a matrix is illustrated in Fig. 2.27. Silica particle is spherical, and its surface is covered with white salt and some shallow pits (Fig. 2.27a). Unfortunately, the pore with 8–10 nm average diameter cannot be discriminated under the condition of

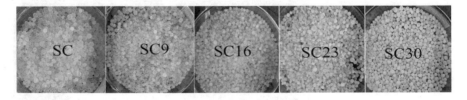

Fig. 2.26 Silica gel samples were impregnated in different concentrations of CaCl$_2$ aqueous solution [59]

(a)   (b)

Fig. 2.27   SEM pictures of type C silica gel after impregnation; **a** 82 times magnification of SC-CaCl$_2$, **b** 30,000 times magnification of SC-CaCl$_2$ [59]

magnification of 30,000 in Fig. 2.27b. Due to the poor electric conductivity of silica gel, it is unable to get more magnification and make the image clearer.

During the impregnation of the material and a series of sorption/desorption cycles, the silica gel ball structure is broken widely in the four samples, which will increase simulation error and uncertainty since it is based on spherical silica sorbents. Meanwhile, the fragment will be more likely dropped out of screen mesh, and it would inevitably lead to the mass decrease of the silica gel, and consequently will impede the mass transfer performance. On the other hand, the carryover problem will occur, and when the salt solution overflows from the matrix, the mass decrease of the salt will decrease the sorption capacity of the sorbent.

Another severe problem is the carryover of the silica gel composite at high humidity and temperature. It occurs when the concentration of the halide solution is too high until the limited pore volume cannot hold all the water sorbed by the compound materials. The phenomenon of carryover depends on concentration or pore volume, as well as the operating condition and experimental duration. Figure 2.28 shows the phenomenon of different extent of carryover of SC30 under the condition of 30 °C and 70% relative humidity with various testing time. In this figure, when the silicone ball transforms from white to transparent, it means that the pore is substantially filled with liquid. When the liquid drops out from the host matrix, the mass of

(a)   (b)   (c)

Fig. 2.28   Different extent of carryover; **a** 80 min, **b** 130 min, **c** 210 min [59]

the CaCl$_2$ in the composite sorbents will decrease, which will have a marked impact on the sorption performance.

References

1. Celzard A, Mareche JF, Furdin G (2005) Modelling of exfoliated graphite. Prog Mater Sci 50:93–179
2. Yue XQ, Yu K, Ji L, Wang ZJ, Zhang FC, Qian LH, Liu YF, Zhang RJ (2011) Effect of heating temperature of expandable graphite on amorphization behavior of powder expanded graphite-Fe mixtures by ball-milling. Powder Technol 211:95–99
3. Dresselhaus MS, Dresselhaus G (2002) Intercalation compounds of graphite. Adv Phys 51(1):1–186
4. Chung DDL (1987) Intercalate vaporization during the exfoliation of graphite intercalated with bromine. Carbon 25(3):361–365
5. Chung DDL (1987) Exfoliation of graphite. J Mater Sci 22(12):4190–4198
6. Tang QW, Wu JH, Sun H, Fang SJ (2009) Crystallization degree change of expanded graphite by milling and annealing. J Alloys Compd 475:429–433
7. Tian B, Yu XG, Wang LW, Wang RZ (2011) Expansion process and thermal conductivity performance of the graphite used as the heat and mass transfer intensification material (in Chinese). J Chem Eng Chin Univ 25(4):572–578
8. Inagaki M, Tashiro R, Washino Y-i, Toyoda M (2004) Exfoliation process of graphite via intercalation compounds with sulfuric acid. J Phys Chem Solids 65:133–137
9. Kang F, Zheng Y, Wang H, Nishi Y, Inagaki M (2002) Effect of preparation conditions on the characteristics of exfoliated graphite. Carbon 40(9):1575–1781
10. Avdeev VV, Martynov IU, Nikolskaya IV, Monyakina LA, Sorokina NE (1996) Investigation of the graphite-H$_2$SO$_4$-gaseous oxidizer (Cl$_2$, O$_3$, SO$_3$) system. J Phys Chem Solids 57:837–840
11. Avdeev VV, Martynov IU, Nikol'skaya IV, Monyakina LA, Sorokina NE (1994) Calorimetric and potentiometry investigations of the acceptor compounds intercalations into graphite. Mol Cryst Liq Cryst 244:115–120
12. Shornikova O, Kogan E, Sorokina N, Avdeev V (2009) The specific surface area and porous structure of graphite materials. Russ J Phys Chem A 83:1022–1025
13. Kang F, Leng Y, Zhang TY (1997) Electrochemical synthesis and characterization of formic acid-graphite intercalation compound. Carbon 35(8):1089–1096
14. Yoshida A, Hishiyama Y, Inagaki M (1991) Exfoliated graphite from various intercalation compounds. Carbon 29(8):1227–1231
15. Inagaki M, Muramatsu K, Maeda Y, Maekawa K (1983) Production of exfoliated graphite from potassium-graphitetetrahydrofuran ternary compounds and its applications. Synth Met 8:335–342
16. Kemin S, Huijuan D (2000) On lower-nitrogen expandable graphite. Mater Res Bull 35:425–430
17. Makotchenko VG, Grayfer ED, Nazarov AS, Kim SJ, Fedorov VE (2011) The synthesis and properties of highly exfoliated graphites from fluorinated graphite intercalation compounds. Carbon 49(10):3233–3241
18. Wei T, Fan Z, Luo G, Zheng C, Xie D (2009) A rapid and efficient method to prepare exfoliated graphite by microwave irradiation. Carbon 47(1):337–339
19. Tryba B, Morawski AW, Inagaki M (2005) Preparation of exfoliated graphite by microwave irradiation. Carbon 43(11):2417–2419
20. Manning TJ, Mitchell M, Stach J, Vickers T (1999) Synthesis of exfoliated graphite from fluorinated graphite using an atmospheric-pressure argon plasma. Carbon 37(7):1159–1164
21. Schlogl R, Boehm HP (1984) The reaction of potassium-graphite intercalation compounds with water. Carbon 22:351–358

22. Skowronski JM (1988) Exfoliation of graphite-CrO_3 intercalation compounds in hydrogen peroxide solution. J Mater Sci 23:2243–2246
23. Dowell MB, Howard RA (1986) Tensile and compressive properties of flexible graphite foils. Carbon 24(3):311–323
24. Gu WT, Zhang W, Li XM, Zhu HW, Wei JQ, Li Z, Shu QK, Wang C, Wang KL, Shen WC, Kang FY, Wu DH (2009) Graphene sheets from worm-like exfoliated graphite. J Mater Chem 19:3367–3369
25. Malas A, Pal P, Das CK (2014) Effect of expanded graphite and modified graphite flakes on the physical and thermo-mechanical properties of styrene butadiene rubber/polybutadiene rubber (SBR/BR) blends. Mater Des 55:664–673
26. Yue XQ, Li L, Zhang RJ, Zhang FC (2009) Effect of expansion temperature of expandable graphite on microstructure evolution of expanded graphite during high-energy ball-milling. Mater Charact 60:1541–1544
27. Han JH, Cho KW, Lee KH, Kim H (1998) Porous graphite matrix for chemical heat pumps. Carbon 36(12):1801–1810
28. Celzard A, Krzesinska M, Mareche JF, Puricelli S (2001) Scalar and vectorial percolation in compressed expanded graphite. Phys A 294:283–294
29. Wang LW, Metcalf SJ, Thorpe R, Critoph RE, Tamainot-Telto Z (2011) Thermal conductivity and permeability of consolidated expanded natural graphite treated with sulphuric acid. Carbon 49(14):4812–4819
30. Wang RZ, Wang LW, Wu JY (2014) Adsorption refrigeration technology: theory and application. Wiley, Singapore
31. Wang LW, Tamainot-Telto Z, Thorpe R, Critoph RE, Metcalf SJ, Wang RZ (2011) Study of thermal conductivity, permeability, and adsorption performance of consolidated composite activated carbon adsorbent for refrigeration. Renew Energy 36:2062–2066
32. Wang LW, Metcalf SJ, Critoph RE, Tamainot-Telto Z, Thorpe R (2013) Two types of natural graphite host matrix for composite activated carbon adsorbents. Appl Therm Eng 50:1652–1657
33. Jin ZQ, Tian B, Wang LW, Wang RZ (2013) Comparison on thermal conductivity and permeability of granular and consolidated activated carbon for refrigeration. Chin J Chem Eng 21(6):676–682
34. Wang LW, Metcalf SJ, Thorpe R, Critoph RE, Tamainot-Telto Z (2012) Development of thermal conductive consolidated activated carbon for adsorption refrigeration. Carbon 50:977–986
35. Zhao YJ, Wang LW, Wang RZ, Ma KQ, Jiang L (2013) Study on consolidated activated carbon: choice of optimal adsorbent for refrigeration application. Int J Heat Mass Transf 67:867–876
36. Zheng X, Wang LW, Wang RZ, Ge TS, Ishugah TF (2014) Thermal conductivity, pore structure and adsorption performance of compact composite silica gel. Int J Heat Mass Transf 68:435–443
37. Eun TH, Song HK, Han JH, Lee KH, Kim JN (2000) Enhancement of heat and mass transfer in silica-expanded graphite composite blocks for adsorption heat pumps: Part I. Characterization of the composite blocks. Int J Refrig 23:64–73
38. Lee CH, Park SH, Choi SH, Kim YS, Kim SH (2005) Characteristics of non-uniform reaction blocks for chemical heat pump. Chem Eng Sci 60:1401–1409
39. Kim ST, Ryu J, Kato Y (2011) Reactivity enhancement of chemical materials used in packed bed reactor of chemical heat pump. Prog Nucl Energy 53:1027–1033
40. Jiang L, Wang LW, Wang RZ (2014) Investigation on thermal conductive consolidated composite $CaCl_2$ for adsorption refrigeration. Int J Therm Sci 81:68–75
41. Zajaczkowski B, Królicki Z, Jezowski A (2010) New type of sorption composite for chemical heat pump and refrigeration systems. Appl Therm Eng 30:1455–1460
42. Han JH, Cho KW, Lee KH, Mauran S (1996) Characterization of graphite-salt blocks in chemical heat pumps. In: Proceedings of international absorption heat pump conference, Montreal, Canada, pp 67–73
43. Mauran S, Coudevylle O, Lu HB (1996) Optimization of porous reactive media for solid sorption heat pumps. In: Proceedings of the international sorption heat pump conference, pp 3–8

44. Wang K, Wu JY, Xia ZZ, Li SL, Wang RZ (2008) Design and performance prediction of a novel double heat pipes type adsorption chiller for fishing boats. Renew Energy 33(4):780–790
45. Oliveira RG, Wang RZ, Wang C (2007) Evaluation of the cooling performance of a consolidated expanded graphite-calcium chloride reactive bed for chemisorption icemaker. Int J Refrig 30(1):103–112
46. Kiplagat JK, Wang RZ, Oliveira RG, Li TX (2010) Lithium chloride—expanded graphite composite sorbent for solar powered ice maker. Sol Energy 84:1587–1594
47. Xu J, Oliveira RG, Wang RZ (2011) Resorption system with simultaneous heat and cold production. Int J Refrig 34:1262–1267
48. Fujioka K, Suzuki H (2013) Thermophysical properties and reaction rate of composite reactant of calcium chloride and expanded graphite. Appl Therm Eng 50:1627–1632
49. Kim ST, Ryu J, Kato Y (2013) Optimization of magnesium hydroxide composite material mixed with expanded graphite and calcium chloride for chemical heat pumps. Appl Therm Eng 50:485–490
50. Py X, Daguerre E, Menard D (2002) Composites of expanded natural graphite and in situ prepared activated carbons. Carbon 40:1255–1265
51. Menard D, Py X, Mazet N (2003) Development of thermally conductive packing for gas separation. Carbon 41:1715–1727
52. Gao J, Wang LW, Wang RZ et al (2017) Solution to the sorption hysteresis by novel compact composite multi-salt sorbents. Appl Therm Eng 111:580–585
53. Wang MZ, He F (1984) Manufacture, property, and application of carbon fiber (in Chinese, ISBN 15030.585). Science Press, Beijing, China
54. Dellero T, Sarmeo D, Touzain P (1999) A chemical heat pump using carbon fibers as additive. Part I: Enhancement of thermal conduction. Appl Therm Eng 19:991–1000
55. Dellero T, Touzain P (1999) A chemical heat pump using carbon fibers as additive. Part II: Study of constraint parameters. Appl Therm Eng 19:1001–1011
56. Vasiliev LL, Mishkinis DA, Antukh AA, Kulakov AG (2004) Resorption heat pump. Appl Therm Eng 24:1893–1903
57. Vasiliev LL, Mishkinis DA, Vasiliev Jr LL (1996) Multi-effect complex compound/ammonia sorption machines. In: Proceedings of international absorption heat pump conference, Montreal, Canada, pp 3–8
58. Vasiliev LL, Mishkinis DA, Antuh A, Snelson K, Vasiliev Jr LL (1999) Multisalt-carbon chemical cooler for space applications. In: Proceedings of international absorption heat pump conference, Munich, Germany pp 579–83
59. Wang JY, Wang RZ, Wang LW (2016) Water vapor sorption performance of ACF-$CaCl_2$, and silica gel-$CaCl_2$, composite adsorbents. Appl Therm Eng 100:893–901
60. Wang LW, Wang RZ, Wu JY, Wang K (2004) Compound adsorbent for adsorptin ice maker on fishing boats. Int J Refrig 27(4):401–408
61. Wang LW (2005) Performances, mechanisms, and application of a new type compound adsorbent for efficient heat pipe type refrigeration driven by waste heat (in Chinese, PhD thesis). Shanghai Jiao Tong University, Shanghai, China
62. Aristov YI, Restuccia G, Caccioba G et al (2002) A family of new working materials for solid sorption air conditioning systems. Appl Therm Eng 22:191–204
63. Tokarev M, Gordeeva L, Romannikov V, Glaznev I, Aristov YI (2002) New composite sorbent $CaCl_2$ in mesopores for sorption cooling/heating. Int J Therm Sci 41:470–474
64. Levitskij EA, Aristov YI, Tokarev MM et al (1996) Chemical heat accumulators: a new approach to accumulating low potential heat. Solar Energy and Solar Cells 44:219–235
65. Restuccia G, Freni A, Vasta S, Aristov YI (2004) Selective water sorbent for solid sorption chiller: experimental results and modeling. Int J Refrig 27:284–293
66. Aristov YI, Tokarev MM, Parmon VN, Restuccia G, Burger HD et al (1999) New working materials for sorption cooling/heating driven by low temperature heat: properties. In: Proceedings of international sorption heat pump conference, Munich, Germany, pp 24–26
67. Daou K (2006) The development, experiment, and simulation of a new type of efficient composite adsorbent driven by the low grade heat source (in Chinese, PhD thesis). Shanghai Jiao Tong University, Shanghai, China

Chapter 3
Properties of Solid Composite Sorbents

Abstract Continuing the contents of Chap. 2, anisotropic thermal conductivity and permeability of consolidated ENG are first presented in this chapter. Based on ENG, thermal conductivity, permeability and sorption capacity of different solid composite sorbents such as AC/ENG, silica gel/ENG, halide/ENG are summarized, following by properties of composite solid sorbents with activated carbon fiber, activated carbon and silica gel.

Keywords Composite solid sorbent · Anisotropic · Thermal conductivity · Permeability · Sorption · Heat transfer · Mass transfer

3.1 Properties of Consolidated ENG

Consolidated ENG (CENG) leads to heterogeneous disordered structure which can be explained by the percolation theory [1, 2]. Celzard et al. studied the vectorial and scalar percolation of CENG, and they found that the CENG appears to behave like a disordered heterogeneous medium where elastic forces are mostly of central nature [3]. Celzard analysed three kinds of CENG as presented in Fig. 3.1. The sample in Fig. 3.1a has low density. The sample in Fig. 3.1b has higher density, and the sample in Fig. 3.1c has the highest density. The experimental results illustrate that the physical property of ENG with low density is isotropic, and the value becomes more and more anisotropic as the density increases [3].

Published papers mostly concentrate on the study of pore structure and elastic properties of CENG. For example, Krzesinska researched the relationships between the parameters for fitting the equations describing the structure and packing geometry of pores of CENG, the elastic modulus of the consolidated matrix and the kind of the raw material [4]. They found that the elastic properties are related to the porous structure of CENG represented by the volume of open and closed pores, specific surface area and pore diameter [5]. As the matrix of the solid sorbents, the ENG is mainly used for improving the permeability and thermal conductivity, thus the anisotropic performance for these two parameters is more important than other characteristics.

© Science Press 2021

L. Wang et al., *Property and Energy Conversion Technology of Solid Composite Sorbents*, Engineering Materials, https://doi.org/10.1007/978-981-33-6088-4_3

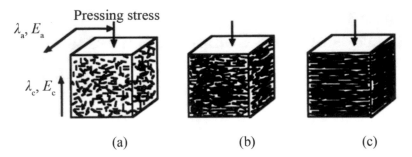

Fig. 3.1 Idealization of cubic samples of CENG samples: definition of two main orthogonal directions along which the thermal conductivity (λ) and the elastic modulus (E) are measured, and visualization of the dependence orientation of graphite sheets versus apparent density of the material. **a** Low density isotropic material; **b** and **c** more and more anisotropic samples [3]

3.1.1   Anisotropic Thermal Conductivity and Permeability of CENG Without GICs

Wang et al. studied the permeability and thermal conductivity of CENG [6]. The ENG is expanded by 700 °C for about 10–15 min. Two moulds (Fig. 3.2) are used for compressing the blocks of ENG. The mould for the disc CENG blocks (DCENG) is shown in Fig. 3.2a, which is mainly for manufacturing the disc blocks in which the permeability and thermal conductivity will be measured parallel to the compressing direction. The steps to produce the blocks involve putting the consolidated sorbent into the vessel, and then compressing it with a force of 0–100 kN applied by the die in a stress force. The thermal conductivity and permeability in the axial direction of the DCENG, which is parallel to the compressing direction, are investigated. The mould for the plate CENG blocks (PCENG) is shown in Fig. 3.2b, which is mainly for producing the blocks for measuring the permeability and thermal conductivity perpendicular to the compressing direction.

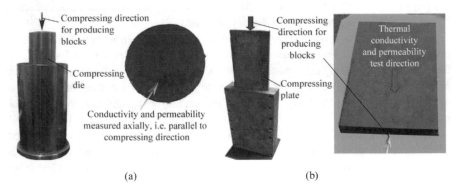

Fig. 3.2   a The mould and sample of DCENG; **b** The mould and sample of PCENG [6]

The thermal conductivity of DCENG (Fig. 3.3a) increases with the increasing density of the disc if the density is lower than 300 kg/m^3. The optimal thermal conductivity is 1.70 W/(m K) for the density of 343–576 kg/m^3. The thermal conductivity declines when the density of the sample is higher than 658 kg/m^3, and the lowest thermal conductivity of 1.21 W/(m K) is obtained when the density is 698 kg/m^3. After that the thermal conductivity increases again with the increasing density. The thermal conductivity of PCENG (Fig. 3.3b) keeps increasing with density. PCENG has a much higher thermal conductivity than DCENG under the condition of similar density, and the difference is larger if the density is higher. For example, the thermal conductivity of PCENG and DCENG is 1.67 and 1.58 W/(m K) respectively when the density is between 210–220 kg/m^3, which illustrates the improvement of PCENG is limited. The thermal conductivity of PCENG and DCENG is 3.13 and 1.40 W/(m K) respectively, when the density is about 660–670 kg/m^3, and the value of PCENG is improved by about 200%.

The anisotropic thermal conductivity is related with the structure of the CENG, for which the SEM pictures (Fig. 3.4) show that kinds of layers of ENG are formed under the stress, which are set perpendicularly to the compressing direction. For the DCENG (Fig. 3.4a), the thermal conductive directions measured are parallel to the compressing direction, i.e. the thermal conductive direction is perpendicular to most micro layers. Such a structure will have larger heat transfer resistance among the layers. For PCENG the thermal conductive direction is perpendicular to the compression direction. Under such a condition since the layers of ENG formed under the pressure are also perpendicular to the compressing direction, the thermal conductive direction is parallel to the distribution of layers (Fig. 3.4b). The distribution of the layers is much more regularly for the samples with higher density, thus the thermal conductivity is also higher.

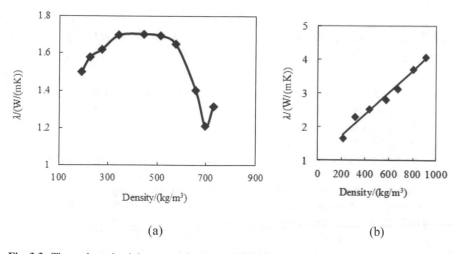

(a)   (b)

Fig. 3.3 Thermal conductivity versus density: **a** DCENG; **b** PCENG [6]

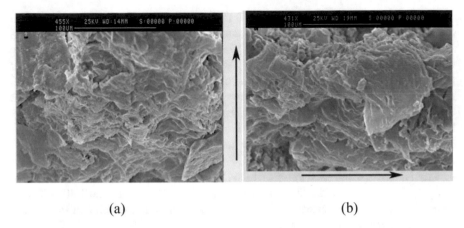

(a)   (b)

Fig. 3.4 SEM pictures (the direction of arrow is the thermal conductive direction): **a** DCENG; **b** PCENG [6]

For the majority of materials, higher thermal conductivity always leads to less gas permeability for same direction, but for the CENG the situation is inversed. The direction that is perpendicular to the compression direction has the capacity of higher thermal conductivity and better gas permeability, which is a kind of phenomenon relevant to the distribution of micro layer. PCENG has a gas transfer and thermal conductive direction parallel to the distribution of micro layer, and this unique structure ensures both lower values of resistance to mass transfer and heat transfer. The permeability of PCENG is shown in Table 3.1, and it ranges from 10^{-12} to 10^{-15} m^2 when the density of PCENG changes from 100 to 500 kg/m^3.

3.1.2 Anisotropic Thermal Conductivity and Permeability of CENG-GICs

Han et al. early studied the anisotropic heat and mass transfer performance of CENG-GICs [8]. The expandable graphite powders used are prepared with HNO_3 as an oxidant and H_2SO_4 as an intercalating additive. They found that the thermal conductivity of a material with graphite matrix is mainly determined by two factors. One is the rotational alignment of graphite basal planes during the compression of ENG powders (Fig. 3.5), and the other is the increase of the contacts between the graphite powders, which decreases the total porosity of the matrix. They got the anisotropy ratio of the thermal conductivity for two directions, i.e. λ_r (the thermal conductivity of a graphite matrix in the radial direction (W/(m K)) and λ_a (the thermal conductivity of a graphite matrix in the axial direction (W/(m K)). With the results of the thermal conductivity shown in Fig. 3.6 and the fact that $\lambda_r/\lambda_a = 200$ at the density of the single graphite crystal (2260 kg/m^3), they got the following empirical expression using polynomial regression:

Table 3.1 Permeability of PCENG [7]

Density (kg/m^3)	100	150	200	250	300	350	400	450	500
Permeability (m^2)	8.914×10^{-12}	2.617×10^{-12}	6.849×10^{-13}	4.367×10^{-13}	1.073×10^{-13}	4.392×10^{-14}	2.923×10^{-14}	1.507×10^{-14}	8.391×10^{-15}

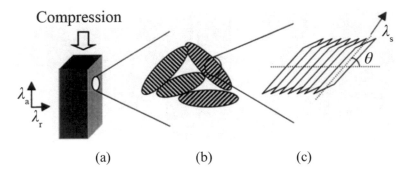

Fig. 3.5 Schematic representation for arrangement of graphite layers: **a** porous graphite matrix for thermal conductivity measurements; **b** ENG-GIC powders; and **c** graphite layers [8]

Fig. 3.6 Anisotropy ratio of graphite matrices versus bulk density of ENG-GIC [8]

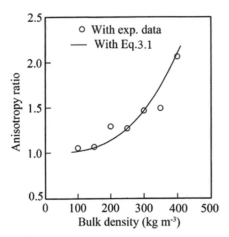

$$\frac{\lambda_r}{\lambda_a} = 1 + \frac{3.247 \times 10^{-2}(\rho_b)^{3.0}}{1.863 \times 10^6 + (\rho_b)} \tag{3.1}$$

where ρ_b is the density of the consolidated matrix.

The anisotropic ratio fitted by Eq. 3.1 (Fig. 3.6) shows that the correlation is satisfactory. The highest anisotropic ratio is about 2 when the density of the matrix is 400 kg/m^3.

After that Bonnissel et al. studied the anisotropy of the ENG-GIC [9]. Since the thermal conductivity of the material depends on the graphite layer orientation, they developed two models defined as a parallel and a perpendicular local thermal conductivity (relative to the compression direction).

$$\lambda_{s\parallel} = \left[\lambda_{M\parallel}f(\theta) + \lambda_{M\perp}(1 - f(\theta)\right]\frac{\rho}{\rho_M} \tag{3.2}$$

$$\lambda_{s\perp} = \left[\lambda_{M\parallel}(1 - f(\theta) + \lambda_{M\perp}f(\theta)\right]\frac{\rho}{\rho_M} \tag{3.3}$$

where $\lambda_{M\parallel}$ (W/(m K)) is the limit parallel thermal conductivity and $\lambda_{M\perp}$ (W/(m K)) is the limit perpendicular thermal conductivity as ρ (local bulk density, kg/m^3) tends toward ρ_M (maximum bulk density, kg/m^3).

$$f(\theta) = 0.5 \text{ for } \rho > \rho_2 \tag{3.4}$$

where ρ_2 is reached when the graphite layers begin to rearrange perpendicular to the compression direction. Below this value the orientation of the graphite layers is random and leads to isotropic properties.

$$f(\theta) = \frac{1}{2}(1 - \cos 2\theta) \text{ for } \rho < \rho_2 \tag{3.5}$$

$$\theta = \frac{\pi}{4}\left(2 - \frac{\rho_M - \rho}{\rho_M - \rho_2}\right) \tag{3.6}$$

Considering a small volume of non-porous graphite around a circular pore of length L_c as presented in Fig. 3.7, they analyzed the anisotropic permeability of the matrix. They assumed that the volume of the graphite crystals remains constant during the compression, but the diameter and the length of the pore change.

For two different directions models are as follows:

$$K_{\parallel} = \frac{l^2}{8\pi} \frac{\left(1 - \frac{\rho}{\rho_M}\right)^2}{\left(c\frac{\rho}{\rho_M} + d\right)^3} \tag{3.7}$$

Fig. 3.7 Elementary model of porosity. Circular pore of length L_c and diameter d_c in the perpendicular direction [9]

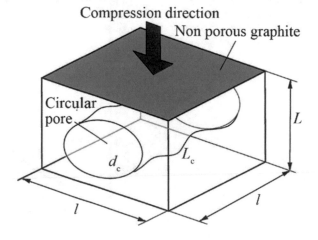

where c and d are empirical constants.

$$K_\perp = \frac{L_M l}{2\pi} \frac{\rho}{\rho_M} \frac{\left(1 - \frac{\rho}{\rho_M}\right)^2}{\left(e\frac{\rho}{\rho_M} + f\right)^3} \qquad (3.8)$$

where e and f are empirical constants, and the subscript of M means maximal.

The overall anisotropy is defined by the ratio of the apparent thermal conductivity.

$$\text{Anisotropy} = \frac{\lambda_{s\perp}}{\lambda_{s\parallel}} \qquad (3.9)$$

The variation of the anisotropy with the overall bulk density computed by the model is shown in Fig. 3.8. The dotted line represents the extrapolated zone above 1400 kg/m^3. The results are much higher than that gotten by Han et al. [8], which are higher than 30 times when the density is about 800 kg/m^3.

The thermal conductivity of the disks of consolidated ENG-TSA (DCENG-TSA) was measured by Wang et al. [10] as the method described for the DCENG. The thermal conductivity of the consolidated samples of disk (Fig. 3.9a) increases rapidly with the increase of density, then stabilises, and decreases finally. The thermal conductivity stabilises at a value of about 5–6 W/(m K) when the density ranges 1000–1200 kg/m^3. The thermal conductivity of the DCENG-TSA shows better performance than that of DCENG, and the highest value for DCENG-TSA is 8.9 W/(m K), which is about 5 times higher than the highest value for DCENG. The plate consolidated ENG-TSA (PCENG-TSA) has much higher thermal conductivity than DCENG-TSA due to the orientation of the micro layers (Fig. 3.9b). Thermal conductivity increases rapidly with the increase of density and the highest thermal conductivity

Fig. 3.8 Variation of the anisotropy with the overall bulk density (kg/m^3). Dotted line represents extrapolated values [9]

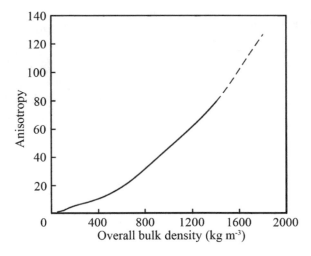

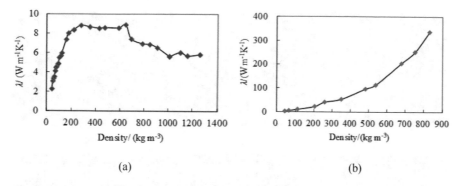

Fig. 3.9 Thermal conductivity versus density: **a** DCENG-TSA; **b** PCENG-TSA [10]

is obtained at a value of 337 W/(m K) when the density of 831 kg/m^3. It reaches nearly one hundred times higher compared to the values of the PCENG. Thus, the corresponding anisotropy is around 50.

The permeability of five PCENG-TSA and five DCENG-TSA samples are listed in Table 3.2. The permeability decreases rapidly with increasing density for both disk and plate samples. The permeability of samples of plate is at least 30 times higher than that of the disk, and the permeability of plate samples is 200 times more than that of disk samples when the density is lower than 200 kg/m^3.

Compared with the consolidated active carbon sorbent LM8 of Ref. [11] for heat pump and refrigeration applications, the thermal conductivity of ENG-TSA is 100 times higher whereas the permeability is similar.

The better performance of PCENG-TSA is credited to the micro layer structure formed by the pressing force. SEM pictures of consolidated samples in Fig. 3.10 show that the micro structure of the samples obtained similar to that of CENG when the density is low, which is a worm structure. But under the condition of

Table 3.2 Permeability of different samples [10]

Type of block	Density (kg/m^3)	Permeability (m^2)	Direction of gas flow
Consolidated samples of disk	75	5.30×10^{-14}	Parallel to the compressive direction
	164	1.12×10^{-15}	
	212	3.89×10^{-16}	
	309	9.15×10^{-17}	
	475	1.49×10^{-17}	
Consolidated samples of plate	111	1.12×10^{-11}	Perpendicular to the compressive direction
	211	1.17×10^{-14}	
	303	2.01×10^{-15}	
	487	3.96×10^{-16}	
	539	1.64×10^{-16}	

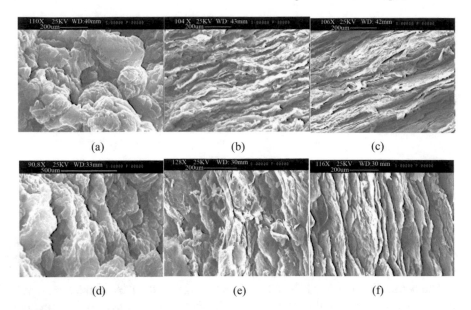

(a)   (b)   (c)

(d)   (e)   (f)

Fig. 3.10 SEM pictures of consolidated samples with different values of density: **a** sample of disk, 65 kg/m³, 110X; **b** sample of disk, 379 kg/m³, 104X; **c** sample of disk, 1084 kg/m³, 106X; **d** sample of plate, 65 kg/m³, 90.8X; **e** sample of plate, 363 kg/m³, 128X; **f** sample of plate, 702 kg/m³, 116X [10]

high density, the worm structure of the samples is replaced by layer structure, which appears to be distributed much more regularly than CENG with similar density. Such a phenomenon accounts for a much higher thermal conductivity of PCENG_TSA compared to PCENG. The SEM figures state probable reasons for the trend of thermal conductivity in Fig. 3.9. For the disk samples (Fig. 3.10a–c), the thermal conductivity is influenced by two factors, which are thermal contact resistance between layers and conduction through the layers, both parallel and perpendicular to the graphite layers. As the disk density increases (up to 250 kg/m³), an initial rise is obtained in thermal conductivity due to the reduction of the thermal contact resistance. With further compression (density 250–650 kg/m³), the thermal conductivity maintains at a stabilized value about 8–9 W/(m K) since the positive influence of reduced resistance between flakes balances with negative influence of layers aligning perpendicular to the heat transfer direction. With almost complete alignment (density above 650 kg/m³), the thermal resistance stabilises at a lower level with all heat transfer in the graphite perpendicular to the plane of the flakes. For the plate samples (Fig. 3.10d–f) the thermal conductivity always increases with the increasing compacting pressure as the heat transfer is all in the plane of the graphite flakes, and the better the alignment is the better the heat transfer is (Fig. 3.9b). Since there is less void space available to allow the passage of gas under the condition of higher density, the permeability shown in Table 3.2 decreases with the density. Therefore, the samples with too high density would be unsuitable for the sorbent matrices due to the poor mass transfer performance.

As explained above, if the intention is to improve thermal conductivity in sorption beds employing a conducting matrix, it is important that the matrix should have higher thermal capacity compared with that of the consolidated sorbent. The specific heat of composite sorbent with two different types of heat transfer matrix, i.e. ENG-TSA and aluminum powder, are compared to confirm the good performance of ENG-TSA using Eq. 3.10, in which C_p is calculated based on 1 kg solid sorbent, i.e. 1 kg AC. This is simply to reflect that the function of the composite material is to sorb. The use of the matrix improves the power density due to the shorter cycle time contributing to the better heat transfer performance. However, from the point of view of thermodynamic cycle efficiency the conductive matrix represents a wasteful thermal mass, i.e., it decreases the efficiency as the specific heat per unit of composite material increases.

$$C_p = \frac{1}{(1-r)} \times \left[(1-r) \times C_{p,AC} + r \times C_{p,\text{matrix}} \right] \qquad (3.10)$$

where C_p is the specific heat capacity for the composite sorbent with 1 kg AC (J/(kg K)), $C_{p,AC}$ and $C_{p,\text{matrix}}$ are the specific heat capacity of AC and heat transfer matrix (J/(kg K)), respectively. r is the ratio of ENG-TSA or aluminum inside the composite sorbents. The $C_{p,\text{matrix}}$ for aluminum is obtained from Ref. [12].

Wang et al. [10] found that the specific heat of ENG-TSA and AC both increase with the increasing temperature and the values of ENG-TSA are much lower than that of AC (Fig. 3.11). The average values of specific heat in the range of 20 to 150 °C are 0.89 and 1.01 kJ/(kg K) for ENG-TSA and AC, respectively, and the value of ENG-TSA is 12% less than the value of AC. The average specific heat of ENG-TSA is slightly lower than the average value of aluminium in the temperature range of 20–150 °C, which is 0.92 kJ/(kg K). The higher thermal conductivity and lower specific heat of compact ENG-TSA indicate that ENG-TSA is a promising

Fig. 3.11 Specific heat versus temperature for ENG-TSA, AC and composite solid sorbents [10]

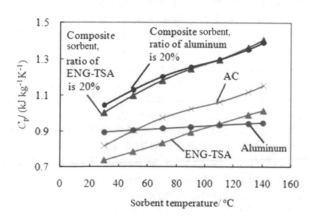

heat transfer matrix, even similar with aluminium. Figure 3.11 also shows that the specific heat of the sample with ENG-TSA as matrix is slightly lower than that of the sample with aluminium as a matrix.

An appropriate figure of the merit of thermal conductivity and specific heat is the thermal diffusivity of the matrix material:

$$\alpha = \lambda/(\rho \times C_p) \tag{3.11}$$

where α is the thermal diffusivity (m^2/s), ρ is density (kg/m^3).

Pure aluminium has a thermal diffusivity of 9.4×10^{-5} m^2/s for the temperature range of 30–150 °C, whereas ENG-TSA conducting perpendicular to the compression direction with a density of 831 kg/m^3 and thermal conductivity of 337 W/(m K) has a thermal diffusivity of 4.6×10^{-4} m^2/s, which is about 5 times higher.

3.2 Properties of Composite Solid Sorbents with ENG

For the consolidated composite sorbents with ENG thermal conductivity always can be optimized by the matrix of the ENG, but the permeability isn't desirable if too high thermal conductivity is pursued. Part of the sorption performance using these kinds of sorbents are analyzed in this section.

3.2.1 The Physical Composite Sorbents

3.2.1.1 AC Composite Sorbents

Jin et al. studied the composite sorbent of AC/ENG. Considering the anisotropic thermal conductivity of sorbents the testing direction they have chosen is the direction that perpendicular to the compressing direction. They chose the sorbent density of 600 kg/m^3 [13], and the weight ratio between AC: ENG is 1:1.5, 1:1, 5:1, 2:1, and 2.5:1 respectively. They tested the thermal conductivity of the sorbents by steady-state heat source method which is based on British Standard BS-874 [11]. The thermal conductivity changes little under testing conditions of different heat power, but it varies considerably with different AC ratios as presented in Fig. 3.12. The highest thermal conductivity is as high as 2.61 W/(m K) when the ratio between AC and ENG is 1:1.5 (AC 40 wt.%), and the value decreases gradually to 2.08 W/(m K) as the AC ratio increases up to 71.4% (2.5:1). The optimum value is improved about 7 times if compared with the results of granular AC and is improved to 2 times if compared with the consolidated sorbent with the chemical binder. But the decreasing permeability happens along with the higher thermal conductivity. For the granular AC with smaller size the permeability gotten from experiments is 4.69×10^{-9} m^2, which

Fig. 3.12 Thermal conductivity of consolidated AC/ENG sorbent versus heating power of central heater [13]

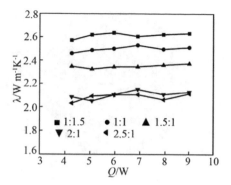

is much higher than the values of consolidated AC/ENG whose value varies from 6.98 × 10^{-13} m^2 to 5.16 × 10^{-11} m^2. Therefore, the composite sorbent of AC with larger density is only suitable for the refrigerant with high pressure, such as ammonia. For the refrigerant works under the condition of vacuum the mass transfer performance caused by the small permeability will influence the sorption performance.

Jin et al. also investigated the thermal conductivity change of composite AC/ENG in sorption process [14], and the results are shown in Fig. 3.13. The density of the composite sorbent increases with the increase of sorption capacity x (kg/kg) (Fig. 3.13a). But because the highest sorption capacity of physical sorbent is only about 0.3 kg/kg, the density doesn't change very much in sorption process. The thermal conductivity of composites almost maintains constant with a variety of sorption capacity (Fig. 3.13b).

Jin et al. tested the permeability for two directions. One is diverging direction for which the refrigerant gas transfers from the inside channel to the outside [14], and

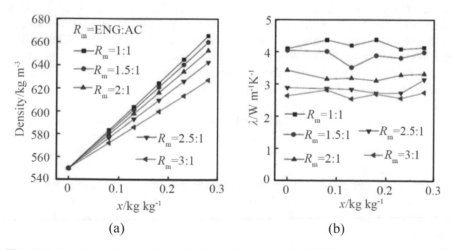

Fig. 3.13 Density and thermal conductivity of composite AC/ENG: **a** density; **b** thermal conductivity [14]

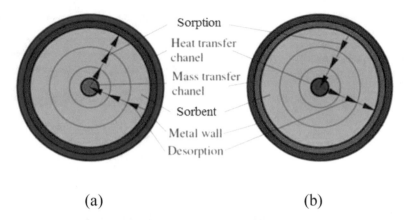

Fig. 3.14 Two types of mass transfer channels for the cylinder beds: **a** diverging direction for sorption and converging direction for desorption; **b** converging direction for sorption and diverging direction for desorption [14]

another is the converging direction for which the refrigerant gas transfers from the outside to the inside channel (Fig. 3.14a, b). Such mass transfer process happens commonly for the cylinder type beds with the mass transfer channel at the center and the heat transfer fluid flowing outside of the cylinder wall. Under such a condition the sorption process is diverging (Fig. 3.14a) for that the gas is transferred from the center channel to outside, whereas the desorption process is converging (Fig. 3.14b) since the gas is transferred from outside to the center mass transfer channel. Before the sorption, the sorbents are tested by the inert gas of N_2.

Results in Table 3.3 show that the permeability decreases while the ratio of AC decreases [14]. The diverging permeability is a bit higher than the converging permeability. The permeability for sorption quantity of 0.28 kg/kg is tested by NH_3, which is shown in Table 3.3. After sorption the converging permeability of AC is similar with that of the experiments with N_2, but the diverging permeability is different. For the large proportion of the AC the converging permeability is slightly larger than that of diverging permeability. The difference between diverging permeability and converging permeability becomes smaller while the proportion of AC decreases. The permeability trend with the fraction of ENG and bulk density is opposite to that of thermal conductivity for composite sorbents. Thus, these two indexes need to be optimized for getting better sorption performances. Table 3.3 also shows that the permeability variation under the condition of different sorption capacity is little and can be neglected.

Biloe et al. investigated the composite sorbent blocks for natural gas sorption by mixing PX-21 type AC and ENG [15], followed by consolidation in a mold. The delivered methane capacity, which is sorbed by sorbent, is close to 100 volumetric methane sorption capacity (V/V). The obtained value of the cycle capacity of methane, i.e. desorbed methane, which is gas, is about 15 V/V. So the total sorption concentration of methane is close to 115 V/V. Thermal conductivity and permeability

Table 3.3 The values of permeability for physical sorbent [14]

Sorption quantity (kg/kg)	Mass transfer mode	Permeability (m^2) (under the condition of $\rho = 550$ kg/m^3)					
		AC:ENG 3:1	AC:ENG 2.5:1	AC:ENG 2:1	AC:ENG 1.5:1	AC:ENG 1:1
0	Converging	5.289×10^{-12}	1.928×10^{-12}	1.827×10^{-12}	6.685×10^{-13}	4.223×10^{-13}
	Diverging	5.839×10^{-12}	2.043×10^{-12}	1.947×10^{-12}	1.024×10^{-12}	3.437×10^{-13}
0.28	Converging	4.671×10^{-12}	7.256×10^{-12}	1.164×10^{-12}	5.285×10^{-13}	3.213×10^{-13}
	Diverging	2.922×10^{-12}	1.134×10^{-12}	9.230×10^{-13}	3.857×10^{-13}	3.200×10^{-13}

are shown in Table 3.4. Permeability and conductivity are measured in the orthogonal direction to the uni-axial direction of compression. Permeability and conductivity increase as weight ratio of ENG increases. Permeability is in the range of around 10^{-13} to 10^{-11} m^2, and it is mainly related with both the geometric flow path and the accessible porosity. ENG improves the thermal conductivity of the composite consolidated blocks to around 2–8 W/(m K), and it is about 10–40 times larger than activated carbon packed bed.

Compared with ENG, the ENG-GICs can optimize the thermal conductivity more significantly. Wang et al. studied the composite AC with the ENG-TSA as the matrix [16]. Wang et al. tested the performance of the composite AC/ENG-TSA, and the samples they chose are the plate samples with the measuring direction perpendicular to the compression direction. The samples they tested are shown in Table 3.5.

The performance of the sorbents is shown in Fig. 3.15. Figure 3.15a shows that for the same bulk density of AC the thermal conductivity always increases though the ratio of AC decreases. Under the condition of same mass ratio, the thermal conductivity increases with increasing bulk density of AC. For some samples, the thermal conductivity decreases if the bulk density of AC is too high due to the cracks happened. The highest thermal conductivity of consolidated composite sorbents reaches the value of 34.15 W/(m K), which is improved by about 150 times if compared with the data of granular AC, i.e. 0.23 W/(m K). The tendency of thermal diffusivity (Fig. 3.15b) is different from the tendency of thermal conductivity since

Table 3.4 Thermal conductivity and permeability of composite AC for methane sorption [15]

Sample no	Density (kg/m^3)	Weight ratio of ENG (%)	Permeability (m^2) $\times$ 10^{13}	Thermal conductivity (W/(m K))
C_0	430	0	–	0.2
C_1	431	10	0.03	1.8
C_2	432	20	0.25	4.4
C_3	475	25	0.9	7.3
C_4	476	30	2.5	6.3

Table 3.5 Parameters of the samples developed for the research [16]

Serial no	Ratio of AC (%)	Grain size of AC (mesh)	Density of sample 1 (kg/m^3)	Density of sample 2 (kg/m^3)	Density of sample 3 (kg/m^3)	Density of sample 4 (kg/m^3)	Density of sample 5 (kg/m^3)
1	33	80–100	278	374	430		
2	40	80–100	215	260	322	419	
3	50	80–100	264	289	302	412	500
4	60	80–100	231	315	401	467	
5	67	80–100	250	308	388	448	
6	71	80–100	190	255	335	372	

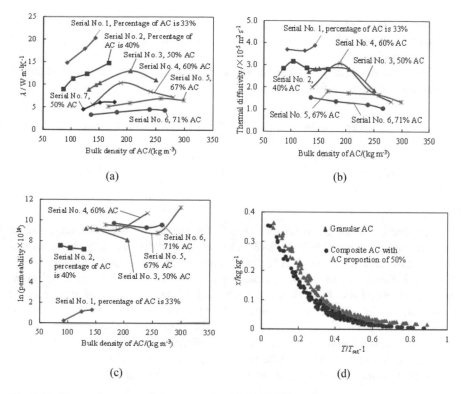

Fig. 3.15 The performance of the composite AC/ENG-TSA sorbents: **a** thermal conductivity; **b** thermal diffusivity; **c** permeability; **d** equilibrium sorption quantity [16]

the thermal capacity and density of sorbents are considered. The optimal thermal diffusivity of different samples is 2.44×10^{-5} m^2/s, which is improved by 45 times if compared with that of granular AC. The permeability (Fig. 3.15c) of consolidated composite AC declines while the ratio of AC in the sample declines. The value is generally equal to or higher than 10^{-11} m^2 as the ratio of AC is larger than 40%, and it declines seriously for the sample with the AC ratio of 33%. The permeability increases slightly when the bulk density of AC increases as the continuous structure of ENG-TSA has been destructed by increasing AC grains in fixed volume, which is beneficial to the mass transfer enhancement. Results also show that the equilibrium sorption performance of consolidated composite AC (Fig. 3.15d) isn't influenced by the matrix of ENG-TSA, and it is similar with the sorption performance of pure granular AC.

For the sorption working pair with AC-ammonia, and the D-A equation [17, 18] is applicable, and it is as follows

$$x = x_0 \exp[-K(\frac{T}{T_s} - 1)^n]$$ (3.12)

where x is the sorption capacity (kg/kg), T is the temperature of sorbent (K), T_s is the saturated temperature of refrigerant (K), x_0 is the maximum sorption capacity, K and n are coefficients.

The performance of AC and composite sorbent of AC is fitted by the exponential equations, which are as follows:

$$x = 0.4655 \times \exp\left[-4.282 \times (\frac{T}{T_{sat}} - 1)^{0.81}\right], R^2 = 0.9698 \text{ Granular AC} \quad (3.13)$$

$$x = 0.4703 \times \exp\left[-5.551 \times (\frac{T}{T_{sat}} - 1)^{0.89}\right], R = 0.9872 \text{ Composite AC}$$

$$(3.14)$$

The coefficient of determination (R^2) for granular AC and composite AC in Eqs. 3.13 and 3.14 are 0.9698 and 0.9872, respectively, and it shows the data obtained from composite AC have slightly higher precision. It is mainly due to the higher heat transfer performance for the composite sorbent. The thermal conductivity of granular AC is much lower than composite sorbent of AC, and the temperature difference between granular AC and the heat source will be higher than that of composite sorbent of AC. Such a phenomenon will cause the slightly higher error between the experimental data and the real data. It should be noted that the error caused by heat transfer is very small and can be neglected. Just as Fig. 3.15 shows, the relative difference between the data of composite AC and granular AC is less than 9%.

3.2.1.2 Silica Gel Composite Sorbents

Eun et al. studied the composite sorbent of silica gel/ENG-GIC, and in ENG-GIC the sulfuric acid is used and the granular expandable graphite is manufactured in China [19]. The consolidate blocks are tested by the direction parallel to the compression direction. It is shown in Fig. 3.16a, and thermal conductivity of composite blocks depends on the graphite weight fraction and the bulk density. Firstly, the heat transfer of composite blocks is enhanced by addition of ENG-GIC (Fig. 3.16a). The effective thermal conductivity of composite blocks, for which the highest value is 19 W/(m K), is much higher than that of silica gel packed bed, 0.17 W/(m K). At constant block density, the conductivity increases with the graphite fraction. With a given graphite fraction the thermal conductivity is larger when the block density is larger. Figure 3.16b shows that the permeability depends on graphite fraction and molding pressure. Lower molding pressure leads to higher permeability. At 8 MPa molding pressure the permeability increases with graphite fraction, and high pressure such as 38 MPa leads to little porosity. Therefore, ENG can help mass transfer in composite block up to certain molding pressure, because it controls the porosity of the block. All points in Fig. 3.16b satisfy the permeability of 10^{-12} m^2, and the permeability will be lower than this data if the pressure is too high. The sorption capacity of silica gel powder and consolidated block of silica gel at equilibrium

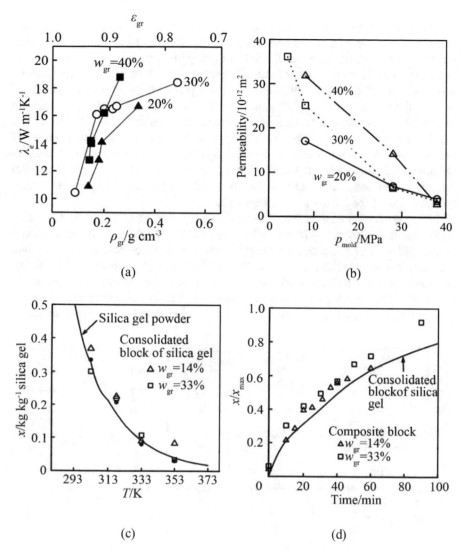

Fig. 3.16 The performance of the composite silica gel/ENG-GIC sorbents, the testing direction is parallel to the compression direction: **a** thermal conductivity versus the bulk density of the graphite; **b** gas permeability of the composite blocks with various graphite weight fractions and molding pressures; **c** sorption isobar of silica gel within blocks when pressure of water is 4.3 kPa, bulk density is 0.80 g/cm^3 for both composites and 0.24 g/cm^3 for a silica gel block; **d** sorption rates of blocks when pressure of water is 4.3 kPa at 30 °C. Bulk density is 0.80 g/cm^3 for composites and 0.24 g/cm^3 for a silica gel block [19]. *Note* w_{gr} means the weight ratio of the ENG-GIC

or under saturated condition are compared, and bulk density and porosity are 0.8 and 0.64 g/cm^3 for relatively dense composite block and 0.24 and 0.89 g/cm^3 for a consolidated block of silica gel. Equilibrium sorption capacity (Fig. 3.16c) of two composite blocks corresponds to that of silica gel powders. By the experimental results they confirmed that ENG-GIC in the composites does not prevent access of water to the surface of silica gel. Figure 3.16d shows sorption rates of the blocks used in Fig. 3.16c. The sorption rate is slightly enhanced by addition of ENG-GIC. Such a result is not due to the improvement of mass transfer but caused by enhancement of heat transfer, for that the composite block (0.8 g/cm^3) is much denser than the silica-gel block (0.24 g/cm^3). For the application of such type of sorbents, they proved that the good mass and heat transfer of the composite blocks can lead to performance enhancement of the refrigeration systems [20].

Zheng et al. [21] studied the pore structure, thermal conductivity and sorption performance of compact composite silica gel and ENG-TSA by simple mixture and consolidation. The relationship among the thermal conductivity, porosity, weight ratio of silica gel, and bulk density of ENG-TSA is presented in Fig. 3.17a–c. The diffusivity and thermal conductivity of the granular silica gel are 0.07 W/(m K) and 5.85 × 10^{-5} kJ/(kg K). The performance of consolidated composite sorbents (Fig. 3.17a, b) reveal that thermal conductivity and diffusivity of composite blocks

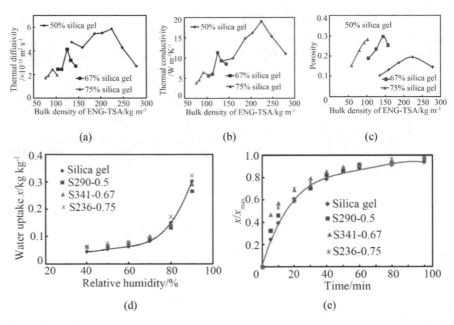

Fig. 3.17 Performances of composite silica gel and ENG-TSA, the testing direction perpendicular to the compression direction: **a** thermal diffusivity; **b** thermal conductivity; **c** permeability; **d** equilibrium sorption performance; **e** sorption rate. *Note* S290-0.5 means that the density of the sample is 290 kg/m^3, and the ratio of the silica gel is 50%. q_t is the water uptake on silica gel at time, kg/kg, and q_∞ means the water uptake on silica gel at equilibrium, kg/kg [21]

are improved considerably by the addition of ENG-TSA. Diffusivity and thermal conductivity of composite blocks are intimately related with bulk density of ENG-TSA and percentage of silica gel. The trends of thermal diffusivity are identical to those of thermal conductivity. The maximum thermal diffusivity is 220 times higher than that of granular silica gel. At the same mass ratio of silica gel, thermal conductivity initially increases sharply with the increasing bulk density of ENG-TSA because of the significantly enhanced thermal conductivity. But after a certain value the thermal conductivity decreases because of increased bulk density of silica gel, i.e. grain separation of silica gel, which blocks up the conduction paths of ENG-TSA. The largest thermal conductivity of consolidated samples is 19.1 W/(m K), which is enhanced by more than 270 times as compared to that of pure silica gel. The porosity (Fig. 3.17c) is higher when the percentage of silica gel is larger. Generally, porosity has a maximum value when the bulk density of ENG-TSA increases. They analyzed the influence of relative humidity on water uptake as shown in Fig. 3.17d, and the equilibrium sorption capacity on both pure silica gel and the other three composite blocks are similar. Consequently, similar to the results of Eun et al. [19], it is confirmed that ENG-TSA in the composites does not affect access of water vapor to the surface of silica gel. Figure 3.17e shows sorption rates of the blocks used in Fig. 3.17d at a relative humidity of 70% and a temperature of 30 °C. At the beginning (the first 10 min) the sorption rate of the blocks with the silica gel ratio of 67% is about twice as much as that of silica gel because of the heat transfer improvement by ENG-TSA. But as the process progresses, the sorption rate of composite silica gel/ENG-TSA starts to decrease slightly faster than that of silica gel until saturation is attained. Such a result indicates that if the composite sorbents are used for sorption desiccant cooling system, the regeneration temperature can be lowered in the same cycle time because of the enhanced sorption rate.

Comparing the data in Figs. 3.16 and 3.17, the thermal conductivity in two figures is similar, but the sorption rate is quite different. For the sample that has the heat and mass transfer direction perpendicular to the compression direction the sorption rate is higher than that parallel to the compression direction.

3.2.2 The Chemical Composite Sorbents

3.2.2.1 ENG Matrix Without GICs

The improvement of the thermal conductivity by the ENG matrix without GICs is limited. For example, Shim et al. studied the MgH_2 compacts containing ENG [22] for hydrogen sorption, and they found the micro layers formed under the pressure as shown in Fig. 3.18a. Then they chose the direction with the optimal heat transfer, i.e. the direction perpendicular to the compression direction, and tested the thermal conductivity of the material. Figure 3.18b shows the results for MgH_2 compacts containing the ENG flakes with an average size of 200 mm as a function of the ENG content measured under hydrogen pressure of 1 bar. The effective thermal

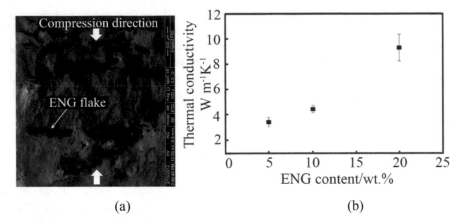

(a)   (b)

Fig. 3.18 Performances of composite MgH$_2$ and ENG, the testing direction is perpendicular to the compression direction: **a** SEM micrograph of the cross-section of the MgH$_2$ compact containing 5 wt.% ENG with an average flake size of 200 μm; **b** effective thermal conductivity of the MgH$_2$ compacts containing the ENG flakes with an average size of 200 μm as a function of the ENG content measured under hydrogen pressure of 1 bar [22]

conductivity is clearly increased with increasing ENG content. While it increases slowly with increasing ENG content up to 10 wt.%, it exhibits a quite high value at 20 wt.% ENG (approximately 9 W/(m K)).

Wang et al. studied the composite sorbents of CaCl$_2$/ENG [23, 24]. They measured the thermal conductivity of the consolidated composite sorbent of 80% CaCl$_2$ and 20% ENG, and the results are shown in Fig. 3.19a. The thermal conductivity increases from 6.5 to 9.8 W/(m K) when the sample molding pressure ranges from 5 to 15 MPa, and corresponding bulk density of sorbent is higher than 1000 kg/m^3. They compared the performance of the consolidated sorbent with the pure granular CaCl$_2$ by the volume cooling density, which is defined as:

$$Q_s = \frac{\Delta x m h_T}{V_a} \tag{3.15}$$

where Δx(kg/kg) is the cycle sorption capacity at a fixed saturated temperature, m is the mass of sorbent (kg); h_T (kJ/kg) is the latent heat of vaporization at that evaporation temperature and V_a (m^3) is the volume of the sorbent.

In comparison with the granular CaCl$_2$, Fig. 3.19b shows that Q_s of composite sorbent is about 44% higher at the evaporating temperature of 0 °C, and it is about 49% higher at the evaporating temperature of -10 °C. For the refrigeration application of such type sorbent, the experimental results of Oliveira et al. show that the specific cooling power and cooling power density can be higher than 1000 W/kg salt and 290 kW/m^3, respectively. These results are obtained at the ambient temperature of 20–30 °C. A kind of reactor using this consolidated sorbent produces a SCP of 415 and 255 W/kg at the average evaporation temperatures of -2.7 and -18.3 °C,

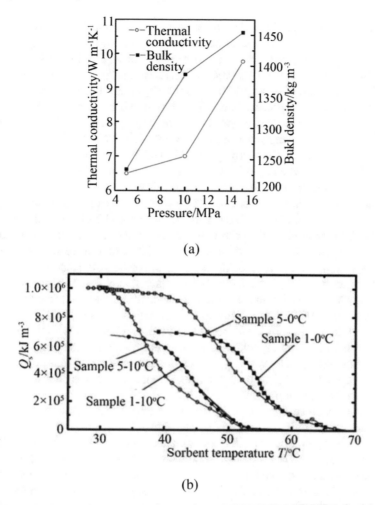

Fig. 3.19 The performance of consolidated composite sorbent of CaCl$_2$/ENG: **a** thermal conductivity and density versus pressure; **b** volume cooling density versus sorbent temperature [23]

respectively. The assessed COP ranges from 0.36 to 0.46 when the global conversion is equal to or higher than 0.5 [25].

Zajaczkowski et al. [26] got the thermal conductivity of 13–15 W/(m K) for the composite sorbent with CaCl$_2$, ENG and CF at the molding force of 500–900 kN, and this value is almost the limitation that the matrix of ENG without GIC can obtain. Huang et al. [27] indicated that the pursue of higher values of thermal conductivity than 14–15 W/(m K) will not significantly improve sorption system efficiency because the mass transfer will be significantly influenced. Such analysis has been performed by Mazet et al. [28]. They obtained higher values of thermal conductivity at the cost of significantly reduced permeability.

The thermal conductivity and permeability of the chemical sorbents will change in the sorption/desorption processes because the serious swelling and agglomeration phenomenon happens. For a powder bed of pure $CaCl_2$, Hosatte and Rheault [29] reported a thermal conductivity of 0.065 W/(m K) for $CaCl_2$-$2NH_3$, of 0.125 W/(m K) for $CaCl_2$-$4NH_3$, and of 0.145 W/(m K) for $CaCl_2$-$8NH_3$. For the system of $CaCl_2$-$8NH_3$ mixed with ENG powder Valkov [30] reported thermal conductivity of 0.49–0.50 W/(m K) when the mass proportion of the ENG (defined as f_g) is 12% and of 0.67–0.74 W/(m K) when f_g is 25%. For system of the ENG/$CaCl_2$-$4NH_3$ mixture, Valkov reported thermal conductivity of 0.35–0.43 W/(m K) when f_g is 12% and of 0.51–0.55 W/(m K) when f_g is 25%. The thermal conductivity of ENG composite sorbent is higher than that of the ammoniated salt alone.

Fujioka and Suzuki [31] studied the effective thermal conductivity of the composite reactant of $CaCl_2$-$3H_2O$ with the matrix of ENG. They firstly soaked and rinsed ENG in ethanol, then rinsed it with water and soaked it in $CaCl_2$ aqueous solution. Lastly, they heated the ENG at around 150 °C to remove excess water until the moles of H_2O in 1 mol of $CaCl_2$ is about 3. At the void fraction of 0.2 to 0.4 for composite the thermal conductivity they have gotten is 10–30 times than that of granular $CaCl_2$ bed that has the thermal conductivity and void fraction of about 0.1 W/(m K) and 0.6. The data increase sharply with decreasing void fraction. They also studied the permeability and found that the Darcy number with the magnitude of 10^{-6} m^2 increases rapidly with decreasing void fraction in the range of 0.4–0.6. Because they have chosen water as the refrigerant, they found that even when the cavitation rate was as high as 0.95, the overall reaction rate was controlled by mass transfer rather than heat transfer.

Jiang et al. studied the composite chlorides with ENG matrix, and they tested the thermal conductivity of the sorbents with different sorption capacity [32]. The thermal conductivity of eight different compact composite ammoniated halides is presented in Fig. 3.20. Take chemical sorbent $SrCl_2$-ENG as an example. Figure 3.20b shows that thermal conductivity is different when the sorption concentration is different, and it increases rapidly with the increasing sorbent concentration in the range of 0–4 mol/mol, and then the trend becomes slowly for the sorption concentration of 5–8 mol/mol. The highest thermal conductivity of composite ammoniated $SrCl_2$ is 2.36 W/(m K), which increases by 182% at most if compared with the data before sorption. For different salts the thermal conductivity ranges between 0.89 and 2.98 W/(m K). The thermal conductivity is also analyzed based on specific volume of salt. The salt with smaller specific volume, such as $FeCl_3$/ENG has higher thermal conductivity (Fig. 3.20a) possibly because of the small thermal resistance between ENG and the surface of halide molecules. The thermal conductivity of $SrCl_2$/ENG and $CaCl_2$/ENG (Fig. 3.20b) is compared, and the data for composite ammoniated $CaCl_2$ with smaller specific volume show the better performance. The sorption capacity of 0–3 mol is selected as a reference to compare the performance of $BaCl_2$/ENG, NH_4Cl/ENG and NaBr/ENG. Figure 3.20c shows that NH_4Cl/ENG, in which NH_4Cl has the smallest specific volume, has highest thermal conductivity. Simultaneously, $BaCl_2$/ENG with the largest specific volume shows the lowest thermal conductivity.

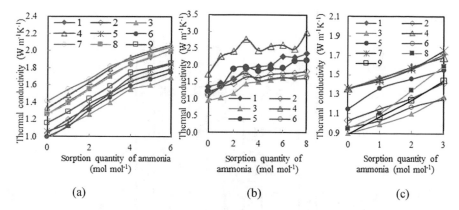

Fig. 3.20 Thermal conductivity of halides/ENG composite versus sorption capacity: **a** sorbents with high desorption temperature and at the initial density of 550 kg/m³, 1-NiCl₂, 75%; 2-NiCl₂, 80%; 3-NiCl₂, 83%; 4-MnCl₂, 75%; 5-MnCl₂, 80%; 6-MnCl₂, 83%; 7-FeCl₃, 75%; 8-FeCl₃, 80%; 9-FeCl₃, 83%; **b** sorbents with middle desorption temperature and at the initial density of 550 kg/m³, 1-SrCl₂, 75%; 2-SrCl₂,80%; 3-SrCl₂, 83%; 4-CaCl₂, 75%; 5-CaCl₂, 80%; 6-CaCl₂, 83%; **c** sorbents with low desorption temperature and at the initial density of 550 kg/m³, 1-BaCl₂, 75%; 2-BaCl₂, 80%; 3-BaCl₂, 83%; 4-NH₄Cl, 75%; 5-NH₄Cl, 80%; 6-NH₄Cl, 83%; 7-NaBr, 75%; 8-NaBr, 80%; 9-NaBr, 83% [32]

Jiang et al. studied the permeability for both converging and diverging modes of eight different compact sorbents [32], and the value ranges from 10^{-14} to 10^{-10} m². The data for the samples of SrCl₂ and CaCl₂ are listed in Table 3.6. The permeability for both converging and diverging ways decrease with increasing sorption capacity. The values for both diverging and converging directions are similar.

3.2.2.2 ENG-GICs Matrix

Mauran et al. proposed a special apparent density for evaluation of the thermal conductivity and permeability [33], which is

$$\rho_{ex} = \frac{m_1}{V_R - V_{sx}} \tag{3.16}$$

where m_1 is the mass of the ENG, V_R is the total volume inside the sorption bed is confined, and V_{sx} is the real volume (at zero porosity) occupied by the halide at the stage of cycle reaction ratio of δx (from 0 to 1).

The experimental results show that for any given set of halide, graphite, and mixing technique, there exists a simple positive linear relationship between the effective thermal conductivity and ρ_{ex} of sorbent, and the relationship for the sorption process of MnCl₂ and NH₃ is shown in Fig. 3.21. It seems the difference between the complex

Table 3.6 Permeability of different compact sorbents [32]

Salts	Density (kg/m³)	Sorption quantity (mol/mol)	State	Percentage of salts		
				83%	80%	75%
SrCl$_2$/ENG	550	0	N$_2$_converging	7.04×10^{-12}	8.69×10^{-13}	4.31×10^{-13}
		8	N$_2$_diverging	3.41×10^{-12}	5.05×10^{-13}	1.77×10^{-13}
		0	NH$_3$_converging	1.15×10^{-12}	8.72×10^{-13}	6.99×10^{-14}
		8	NH$_3$_diverging	3.47×10^{-13}	2.32×10^{-13}	1.47×10^{-14}
CaCl$_2$/ENG	550	0	N$_2$_converging	9.06×10^{-12}	5.17×10^{-12}	2.18×10^{-12}
		8	N$_2$_diverging	3.99×10^{-12}	3.55×10^{-12}	3.29×10^{-12}
		0	NH$_3$_converging	1.57×10^{-13}	2.50×10^{-13}	1.99×10^{-13}
		8	NH$_3$_diverging	2.27×10^{-13}	2.33×10^{-13}	1.45×10^{-13}

Fig. 3.21 Evolution of the
effective thermal
conductivity of IMPEX as a
function of ρ_{ex} for
anhydrous compositions [33]

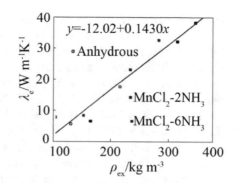

of $MnCl_2$-$2NH_3$ and $MnCl_2$-$6NH_3$ isn't big. The thermal conductivity of IMPEX is
as high as about 38 W/(m K) when ρ_{ex} is around 350 kg/m^3.

Mazet et al. studied the IMPEX as well, and they experimentally obtained
thermal conductivity of 0.2–40 W/(m K) [34]. They also [34, 35] analyzed thermal
conductivity of IMPEX blocks with apparent density of 200 kg/m^3 and obtained
16 W/(m K).

Han et al. studied the expandable graphite with a small amount of stage-5 structure
[36], which is prepared with H_2SO_4 as an intercalating additive and HNO_3 as an
oxidant. They studied the gas permeability of the composites by Darcy's law [37].
The reactor is cylindrical. The gas permeability is measured with the composite in a
direction perpendicular to the compression direction, which corresponds to the radial
direction of the reactive composite of cylindrical shape. They studied the permeability
change in the sorption/desorption phase. The results of different salts are shown in
Fig. 3.22, in which f_g is the mass proportion of the ENG-GIC in the composite, and
bulk density ρ_b is for the bulk density of ENG-GIC in the composite (the mass of

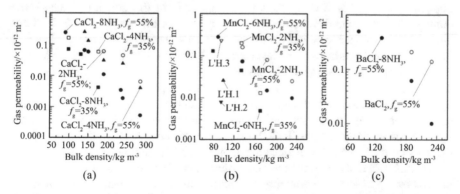

Fig. 3.22 The permeability of different sorbents: **a** Gas permeability of graphite/$CaCl_2$-nNH_3
composites; **b** Gas permeability of graphite/$MnCl_2$-nNH_3 composites; L'H 1: measured by
L'Haridon et al. for $MnCl_2$-$6NH_3$ when $f_g = 35\%$; L'H 2: measured by L'Haridon et al. for
$MnCl_2$-$6NH_3$ when $f_g = 25\%$; L'H 3: measured by L'Haridon et al. for $MnCl_2$-$2NH_3$ when f_g
$= 25\%$; **c** Gas permeability of graphite/$BaCl_2$-nNH_3 composites [36]

ENG-GIC/the volume of the composite). Figure 3.22a shows the permeability of the ENG-GIC/CaCl$_2$-nNH$_3$ composites, and the data is in the range of 5.3×10^{-16} to 10^{-12} m^2 with ρ_b of 80–300 kg/m^3. As ρ_b increases, the gas permeability reduces significantly because of the increase of mass transfer resistance by the compression. Furthermore, at a fixed ρ_b and f_g of ENG-GIC/CaCl$_2$-nNH$_3$ composites, the gas permeability decreases 5–22 times with the increase of n. Figure 3.22b shows the gas permeability of the ENG/MnCl$_2$-nNH$_3$ composites. The gas permeability is in the range of 4.9×10^{-15} to 10^{-12} m^2 with ρ_b of 80–250 kg/m^3. They compared the data with that gotten by L'Haridon and Mauran for graphite/MnCl$_2$-6NH$_3$ composite with ρ_b of 100 kg/m^3 and $f_g = 35\%$, and values are similar. Figure 3.22c shows the gas permeability of ENG-GIC/BaCl$_2$-nNH$_3$ composite. The gas permeability is in the range of 1.1×10^{-14} to 10^{-12} m^2 with ρ_b of 90–230 kg/m^3. The overall behavior of the gas permeability for graphite/MnCl$_2$-nNH$_3$ and BaCl$_2$-nNH$_3$ composites is similar to the case of graphite/CaCl$_2$-nNH$_3$ composite.

Han et al. also studied the thermal conductivity of different halides in sorption or desorption processes [38], and the values are distinctively high. The thermal conductivity of the graphite/CaCl$_2$-nNH$_3$ complex is shown in Table 3.7. When f_g is 70% the thermal conductivity of graphite/CaCl$_2$-2NH$_3$, graphite/CaCl$_2$-4NH$_3$, and graphite/CaCl$_2$-8NH$_3$ complex is 18.0–35.3 W/(m K), 21.2–40.2 W/(m K), and 24.0–48.6 W/(m K) respectively with bulk density of 110–380 kg/m^3. The thermal conductivity depends on the ammoniated state that may change the total porosity in the complex. When f_g is 50%, the thermal conductivity of the graphite/CaCl$_2$-2NH$_3$ complex is 15.1–25.0 W/(m K) with bulk density of 110–340 kg/m^3. The thermal conductivity of the graphite/CaCl$_2$-4NH$_3$ complex and the graphite/CaCl$_2$-8NH$_3$ complex are in the range of 17.1–28.2 and of 20.0–32.5 W/(m K), respectively. Such a result indicates that the effective thermal conductivity decreases as the weight fraction of the halide increases.

Table 3.8 shows the effective thermal conductivity of the graphite/MnCl$_2$-nNH$_3$ complex. When f_g is 50% the thermal conductivity of the graphite/MnCl$_2$-2NH$_3$ complex and the graphite/MnCl$_2$-6NH$_3$ complex is in the range of 13.7–21.8 and 14.6–22.9 W/(m K), respectively, with the bulk density of 95–235 kg/m^3. When f_g is 30%, the thermal conductivity of the graphite/MnCl$_2$-2NH$_3$ and the graphite/MnCl$_2$-6NH$_3$ complex is in the range of 10.8–12.7 and 11.5–14.1 W/(m K) separately, at bulk density of 90–145 kg/m^3. Han et al. tried to compare the data with that gotten by Mauran et al. [33], unfortunately they didn't get the distinct results since Mauran et al. did not designate the weight fraction in their work. The highest value of the thermal conductivity for MnCl$_2$-6NH$_3$ gotten by Mauran et al. is 38.1 W/(m K) at the density of 350 kg/m^3. But since the apparent density defined by Mauran et al. is the volume difference between the total volume and the volume occupied by the salt divided by the mass of the graphite, the results cannot be compared each other.

Table 3.9 shows the effective thermal conductivity of the graphite/BaCl$_2$-8NH$_3$ complex. When f_g is 50%, thermal conductivity of the complex is in the range of 16.9–30.7 W/(m K) with bulk density of 100–255 kg/m^3. When f_g is 30% thermal conductivity is in the range of 9.6–16.2 W/(m K) with bulk density of 75–160 kg/m^3.

Table 3.7 Effective thermal conductivity of graphite/$CaCl_2$-nNH_3 complex [38]

ρ_b (kg/m^3)	f_g (wt.%)	n (for $CaCl_2$-nNH_3)	Thermal conductivity (W/(m K))
110	70	8	24
		4	21.2
		2	18
290	70	8	38
		4	31.9
		2	29.3
380	70	8	48.6
		4	40.2
		2	35.3
110	50	8	20
		4	17.1
		2	15.1
230	50	8	28.3
		4	22.6
		2	21
340	50	8	32.5
		4	28.2
		2	25
105	30	8	14.1
		4	11.3
		2	10.2
165	30	8	16.1
		4	13.8
		2	12.9

The overall behavior of the effective thermal conductivity is similar to the case of the graphite/$CaCl_2$-nNH_3 and the graphite/$MnCl_2$-nNH_3 complexes.

Bou et al. [39] introduced layered flake graphite blocks, characterized each graphite block with different apparent density, and dispersed the active agent in it. The reaction rate is significantly increased and the reaction time is equal to or less than 40 min.

Jiang et al. studied the composite chlorides with ENG-TSA as matrix [40]. They studied 35 types of samples, and the detailed composition is shown in Table 3.10. Because the refrigerant for the composite sorbents is ammonia and the working pressure is positive, the density of 300–600 kg/m^3 is chosen. The percentage of halide in the composite sorbent is 50%, 67%, 75%, 80% and 83% respectively. The bulk density of pure halide in consolidated sorbent is calculated by dividing the whole volume with the mass of sorbent, as shown in Table 3.10.

Table 3.8 Effective thermal conductivity of graphite/$MnCl_2$-nNH_3 complex [38]

ρ_b (kg/m^3)	f_g (wt.%)	n (for $MnCl_2$-nNH_3)	Thermal conductivity (W/(m K))
95	50	6	14.6
		2	13.7
155	50	6	16.4
		2	15.7
235	50	6	22.9
		2	21.8
90	30	6	11.5
		2	10.8
120	30	6	13.2
		2	11.5
145	50	6	14.1
		2	12.7

Table 3.9 Effective thermal conductivity of graphite/$BaCl_2$-$8NH_3$ complex [38]

ρ_b (kg/m^3)	f_g (wt.%)	Thermal conductivity (W/(m K))
100	50	16.9
170	50	21.5
255	50	30.7
75	30	9.63
160	30	16.2

Table 3.10 Bulk density of halide versus density of composite sorbent [40]

Serial, No	Mass ratio of $CaCl_2$(%)	Density (kg/m^3) of different samples						
		Sample 1	2	3	4	5	6	7
1	50	150	175	200	225	250	275	300
2	67	200	233.3	266. 7	300	333.3	366.7	400
3	75	225	262.5	300	337.5	375	412.5	450
4	80	240	280	320	360	400	440	480
5	83	250	291. 7	333. 3	375	416.7	458.3	500

Thermal conductivity of consolidated sorbents with different ratios and bulk density is shown in Fig. 3.23. Thermal conductivity ranges from 23.5 to 88.1 W/(m K), and it increases with the increase of bulk density and the decrease of mass ratio of $CaCl_2$. The highest thermal conductivity of sample is 88.1 W/(m K) for Sample 1 of Serial No. 1. Compared with $CaCl_2$ impregnated with ENG [41] the optimal thermal conductivity of $CaCl_2$ impregnated with ENG-TSA is about 44 times higher, and it is 440 times higher than that of granular $CaCl_2$.

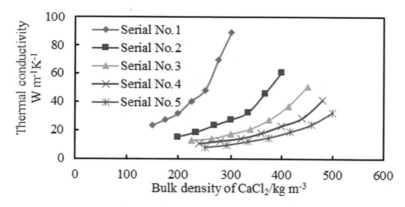

Fig. 3.23 Thermal conductivity versus mass ratio and bulk density of CaCl₂ [40]

The permeability of different sorbents is presented in Fig. 3.24. The permeability increases when the bulk density decreases and mass ratio of salt increases. For various bulk density and mass ratios, the permeability of consolidated sorbent ranges from 9.31×10^{-10} to 3.05×10^{-14} m². When the permeability is very low, i.e. for the CaCl₂ percentage of 50 and 67%, the permeability decreases slightly when the bulk density of CaCl₂ increases. When the percentage of CaCl₂ is larger than 67%, the permeability of composite sorbent decreases significantly. With increasing bulk density of CaCl₂, the permeability increases because the CaCl₂ cannot be compressed very much and it resists the compression of the ENG-TSA and establishes its continuous structure for gas permeation.

Jiang et al. also studied the permeability of pure CaCl₂ in sorption process when the expansion space is about 3–4 times of the volume for CaCl₂. Results are listed in Table 3.11.

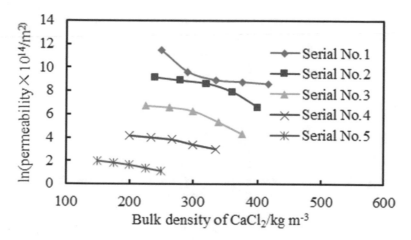

Fig. 3.24 Permeability of CaCl₂ consolidated sorbent [40]

Table 3.11 Values of permeability for granular $CaCl_2$ in sorption process [40]

Sorption capacity (mol/mol)	0	1	2	3	4	5	6	7	8
Permeability (m^2)	2.34×10^{-9}	7.78×10^{-15}	6.55×10^{-15}	9.26×10^{-16}	8.18×10^{-16}	5.87×10^{-16}	5.78×10^{-16}	3.65×10^{-16}	1.56×10^{-16}

Table 3.11 shows that granular $CaCl_2$ has very high permeability before it sorbs ammonia because the void among the granules of $CaCl_2$ provides the abundant mass transfer channels. But after sorption, the agglomeration happens and the mass transfer channels are blocked, consequently the permeability of the pure $CaCl_2$ decreases observably, which decreases from 2.34×10^{-9} to 1.56×10^{-16} m^2. Compared the data in Table 3.11 with the data in Fig. 3.24, the consolidated composite sorbent has lower permeability than that of granular $CaCl_2$ before the sorption, but it has much higher permeability after sorption. At different bulk densities and mass ratios, the permeability of consolidated composite sorbent ranges from 9.31×10^{-10} to 3.05×10^{-14} m^2, and it is improved by higher than 100 times if compared with the data for the granular $CaCl_2$ after sorption. It is mainly because the porous ENG-TSA provides abundant mass transfer channels for the sorbent after sorption, which improves the permeability effectively. Such a result indicates that the serious performance attenuation may happen for the refrigerator of pure $CaCl_2$ because of the reduced permeability caused by agglomeration, and the application of composite sorbent with ENG matrix may be a solution for that.

3.2.2.3   Non-equilibrium Sorption Performances for Chemical Composite Sorbents with ENG-TSA

ENG-TSA was used as the inert porous additive of the halides. The test unit of sorption performance can be seen in [42]. First, Clapeyron curves of $MnCl_2$/ENG-TSA-NH_3 and $CaCl_2$/ENG-TSA-NH_3 under the non-equilibrium conditions are tested and the results are presented in Fig. 3.25.

Figure 3.25 shows that the non-equilibrium Clapeyron line of the experiment is parallel to the theoretical line, but very different from the theoretical line. More ammonia gas sorbed results in more difference between theoretical and experimental

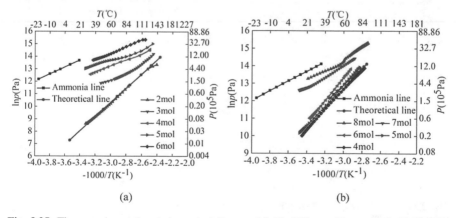

(a)   (b)

Fig. 3.25 The experimental and theoretical lines: **a** $MnCl_2$-NH_3 working pair; **b** $CaCl_2$-NH_3 working pair [42]

results. This result indicates the experimental sorption/desorption performance of
MnCl$_2$ has binary variables of p and T other than single variable of p or T.

The isobaric sorption/desorption performances are shown in Fig. 3.26. It can be
seen from Fig. 3.26a that the actual isobaric sorption curve of MnCl$_2$-NH$_3$ working
pair is in good agreement with the experimental results in Fig. 3.25a, but the isobaric
desorption line is greatly different from the experimental results in Fig. 3.25a. It
means that if the experimental Clapperon curve is used to simulate the refrigeration
performance, the obtained data will have a large error due to the obvious hysteresis
in the actual desorption process. According to Fig. 3.26a, when the heat source
temperature is about 110 °C, MnCl$_2$-NH$_3$ has the largest hysteresis effect. This
data is lower than the equilibrium desorption temperature of 120 °C. The lower
the ammonia content in ammoniated MnCl$_2$, the weaker the hysteresis. When the
sorption capacity is close to 2 mol of NH$_3$ /mol MnCl$_2$, it almost disappears.

The CaCl$_2$-NH$_3$ has similar trends with the MnCl$_2$-NH$_3$ working pair (Fig. 3.26b).
The isobaric sorption curve in Fig. 3.26b is close to the experimental results in
Fig. 3.25b, but the desorption curve is much different. The CaCl$_2$-7NH$_3$ shows
obvious hysteresis. Compared with the results in Fig. 3.25b, the desorption hysteresis
is more serious when the experimental Clapeyron line is far from the theoretical
equilibrium line, i.e. when the sorption capacity is larger.

With the similar procedure, the sorption performance of compact composite bi-
salt and multi-salt sorbents with ENG-TSA are tested [43]. Figure 3.27 shows the
Clapeyron curve for the bi-salt mixture (CaCl$_2$/NH$_4$Cl). The temperature and pressure of bi-salt mixture of NH$_4$Cl/CaCl$_2$ are measured under different sorption capacities, and the Clapeyron curve of single salt is compared. As presented Fig. 3.27, when
the sorption capacity of bi-salt mixture is about 4.6 mol of NH$_3$ /mol of chlorine, the
curve basically coincides with the theoretical Clapeyron curve of NH$_4$Cl. When the
sorption capacity is low, such as when the minimum sorption capacity is 0.7 mol of

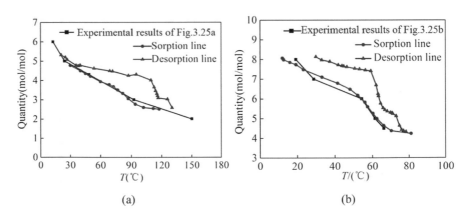

(a)   (b)

Fig. 3.26 Isobaric sorption/desorption curves under the condition of 4.3 bar: **a** MnCl$_2$-NH$_3$;
b CaCl$_2$-NH$_3$ [42]

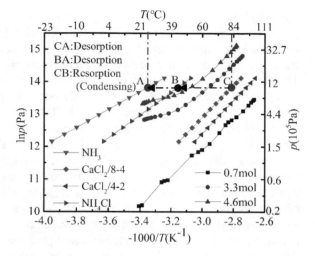

Fig. 3.27 The Clapeyron diagram of bi-salt under non-equilibrium conditions [43]

NH$_3$ /mol halide, the curve is close to the Clapeyron curve of CaCl$_2$/4–2 under the equilibrium condition.

The sorption hysteresis phenomenon for the bi-salt mixture is studied, and the isobaric sorption/desorption performances under the condition of different pressures are shown in Fig. 3.28.

As shown in Fig. 3.28a, for the bi-salt mixture, there is a step from 20 °C to about 58 °C at a pressure of 360 kPa. According to the theoretical Clapeyron curve under equilibrium conditions (as shown in Fig. 3.27), pressure and temperature are one-to-one. Since the pressure is 360 kPa, the level of 360 kPa intersects the theoretical curves of NH$_4$Cl and CaCl$_2$, and the corresponding temperatures are 16 °C and 57 °C, respectively. Therefore, it can be concluded from Fig. 3.28a that the mixture shows the combination of the two halides it contains. A similar conclusion can be drawn from Fig. 3.28b. Figure 3.28a, b show that the sorption hysteresis under different

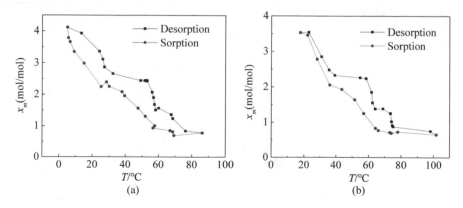

Fig. 3.28 The isobaric sorption/desorption curves of bi-salt: **a** 360 kPa, **b** 430 kPa [43]

pressures is similar and is much less significant than that of the single halide sorbent, as shown in Fig. 3.26. Take the points A, B, and C in Fig. 3.27 as an example for analysis. Assuming that NH_3 is desorbed from $CaCl_2$ at a certain pressure and temperature, the desorption point is C and the condensation point is A in Fig. 3.27. The desorption line CA intersects the theoretical curve of NH_4Cl at point B. For bi-salt, three processes may continue. The first is process C $\rightarrow$ A, where NH_3 is desorbed from $CaCl_2$; the second process is B $\rightarrow$ A, where NH_3 is desorbed from NH_4Cl. The last process C $\rightarrow$ B is the most interesting process, it may be the sorption between $CaCl_2$ and NH_4Cl. For the sorption process, the heat of reaction is released to the outside through NH_4Cl. Part of the heat of desorption and heat of sorption can be neutralized, which leads to a decrease in the total heat of reaction. As mentioned by many researchers, since the hysteresis is believed to be related to the reaction heat of the desorption process [17], the reduction may make the sorption hysteresis less severe than that of a single salt. Therefore, it can be concluded that multi-salt may become a solution to the common hysteresis phenomenon of a single halide. The Clapeyron curve of the tri-salt mixture is shown in Fig. 3.29.

Similar to the performance of the bi-salt mixture, the curve of high sorption capacity (for example, 4.9 mol ammonia/mol mixture) coincides with the theoretical Clapeyron curve of NH_4Cl low-temperature halide, and the lowest sorption capacity curve is 1.4 mol NH_3/mol mixture close to that of $MnCl_2$ high-temperature salt theoretical curve. The curve of about 2.6 mol NH_3/mol mixture coincides with the theoretical Clapeyron curve of $CaCl_2$, reflecting the characteristics of mid-temperature salt. As the temperature increases, the distance between the experimental curves will increase. It shows that the less the amount of ammonia composites by the halide, the greater the energy required for desorption, that is, desorption becomes much more difficult. This phenomenon once again shows that the chemical sorption process is bi-variate, rather than uni-variate control.

Figures 3.27 and 3.29 show that the performance of multi-salt sorbents is related to the composites of different halides. Since the various characteristics of halides

Fig. 3.29 The Clapeyron diagram of tri-salt mixture under non-equilibrium conditions [43]

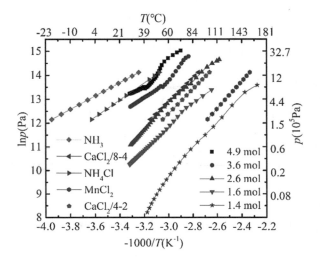

affect each other, and the energy required for desorption may be different from the sum of all the halides it contains, this will have an impact on the sorption hysteresis. Then the isobaric sorption/desorption characteristics at 360 and 430 kPa pressures are studied, and the results are shown in Fig. 3.30.

Three steps are shown clearly in Fig. 3.30a, b. The turning points for different pressure are close to the equilibrium data of different halides. For example, for the working pressure of 360 kPa presented in Fig. 3.29, the equilibrium temperatures for Clapeyron curves are about 16 °C, 57 °C, 69 °C and 116 °C respectively for the halides of $NH_4Cl/3–0$, $CaCl_2/8–4$, $CaCl_2/4–2$ and $MnCl_2/6–2$. The corresponding turning temperatures in Fig. 3.30a are 20 °C, 60 °C and 120 °C separately, which are similar with the data of single-halide.

The multi-halide mixture shown in Fig. 3.30 has a novel phenomenon that no hysteresis occurs this time. As discussed, hysteresis was discovered earlier by Goetz and Marty [44], and then by Wang [45], Trudel et al. [46], Aidoun and Ternan [47] have discussed it. Generally, scholars believe that hysteresis is a universal phenomenon, which occurs in almost all types of chemical sorbents. Figure 3.30 shows the solution to this phenomenon and shows that more salt can make the hysteresis almost disappear. When the pressure changes, there is no hysteresis loop in Fig. 3.30a, b, as previously shown in Figs. 3.26 and 3.28. Especially for the working pressure of 430 kPa, the sorption curve almost overlaps with the desorption curve. Similar to the bi-halide sorbent, it should be related to the reduced heat of reaction. The sorption between NH_4Cl, $CaCl_2$ and $MnCl_2$ may lead to a reduction in the energy difference. The sorption and desorption of tri-halide sorbents will be more complicated, and the energy difference in the sorption and desorption process may change greatly. But each salt maintains its inherent characteristics due to physical mixing.

G. L. An compared the hysteresis characteristics of $CaCl_2$ and multi-halide sorbents [48]. The results show that under equilibrium condition the hysteresis characteristic of $CaCl_2$ does not change inside the multi-halide sorbent, which means the inner resorption does not occur. The non-equilibrium experiments are carried out,

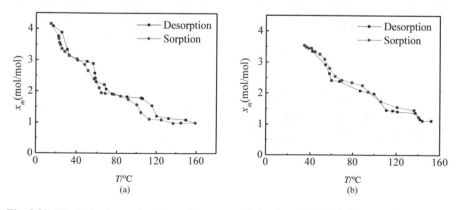

Fig. 3.30 The isobaric sorption/desorption curves of tri-salt: **a** 360 kPa, **b** 430 kPa [43]

including the influence of chemical reaction, heat transfer, and mass transfer. Results show that the multi-halide sorbent could decrease the hysteresis temperature difference significantly from 23.4 to 7.8 °C if compared with $CaCl_2$ when the evaporation pressure is controlled at 0.441 MPa. Taking desorption process as example, due to the existence of heat transfer temperature difference, the temperature of the material near to external r_{out} is higher than that near to inner r_{in}, which causes the reaction rate of sorbent in external layer larger than that in the inner layer. Thus, when temperature is just over threshold temperature of NH_4Cl, only part of $[NH_4(NH_3)_3]Cl$ located at r_{out} begins to desorb. While during stage in Fig. 3.31a, before all of the NH_4Cl sorbents complete desorption process, $[Ca(NH_3)_8]Cl_2$ becomes active to desorb. This duration is called simultaneous reaction stage. The similar process (Fig. 3.31b) exists between middle and high temperature halide sorbents, which means better flexibility and faster response to heterothermic condition due to the multi-stage reaction property. Thus, the continuous reaction process combined with non-equilibrium heat and mass transfer leads to higher sorption temperature and lower desorption temperature of multi-halide sorbent than that of single halide ($CaCl_2$) under the same condition, and such a phenomenon reduce the sorption hysteresis effectively and could adapt well to the solar energy.

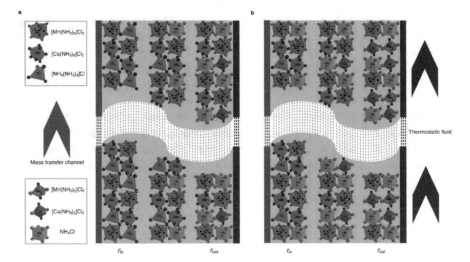

Fig. 3.31 Relationship between spatial position of sorption bed (r) and reaction state of different halides inside multi-halide sorbent, during whole desorption stages under non-equilibrium conditions: **a** Temperature reaches desorption temperature of $CaCl_2$ and two halides (NH_4Cl and $CaCl_2$) react simultaneously, **b** Temperature reaches desorption temperature of $MnCl_2$ and two halides ($CaCl_2$ and $MnCl_2$) react simultaneously [48]

3.3 Properties of Composite Solid Sorbents with Activated Carbon and Activated Carbon Fiber

3.3.1 Composite Solid Sorbents with Activated Carbon

Wang et al. apply activated carbon as matrix to halide-ammonia working pairs, aiming at improving the mass transfer of chemical sorbent by adding activated carbon. Heat transfer will be further improved if the composite is consolidated [49].

The sorption performance of $CaCl_2$ was tested, indicating that the performance of the sorbent is closely related to the retention and expansion space of the sorbent [50, 51]. Due to the agglomeration and swelling phenomenon in the sorption process, serious agglomeration and swelling will cause mass transfer problems and affect the sorption and refrigeration performance when the limited sorbent expands the space.

For the environmental temperature of 30 °C and air conditioning temperature of 9 °C, $CaCl_2$ is tested and the optimal filling volume is only 41.6% in the sorption bed. If filling capacity quantity is more than 41.6%, sorption performance of the sorbent will degrade due to the problem of mass transfer. For the same condition, the results of simply mixed sorbent as well as the consolidated sorbent with different volume ratio are shown in Tables 3.12 and 3.13. Under the condition 3 in Table 3.12, due to the high volume percentage of $CaCl_2$, increment of volume cooling capacity of sorbent is lower than that of volume filling capacity, which shows that agglomeration and swelling influence the mass transfer of sorbent as well as the refrigeration performance of the sorbent.

Table 3.12 Performance of simply mixed composite sorbent for air conditioner (compared with pure $CaCl_2$ sorbent) [52]

	$CaCl_2$ ratio (%)	Activated carbon ratio (%)	Increment of filling volume (%)	Increment of volume cooling capacity (%)
Condition 1	70	30	29	29
Condition 2	79	21	32	32
Condition 3	82	18	35	26

Table 3.13 Performance of consolidated composite sorbent for air conditioner (compared with pure $CaCl_2$ sorbent) [52]

	$CaCl_2$ content (%)	Activated carbon ratio (%)	Water ratio (%)	Cement ratio (%)	Increment of filling volume (%)	Increment of volume cooling capacity (%)
Condition 1	55.3	24	13.8	6.9	29	29
Condition 2	60.2	17.2	15.1	7.5	35	35
Condition 3	57.8	20.5	14.5	7.2	32	32

For the environmental temperature of 28 °C and freezing temperature lower than -10 °C, experiments show that the optimal filling volume of $CaCl_2$ is only 33.3%. If the value is larger than this data, the mass transfer performance will be critical. The performance of simply mixed sorbent as well as consolidated sorbent with different ratio is listed in Tables 3.14 and 3.15. Condition 2 of Table 3.15 is only suitable when refrigeration temperature is -10 °C. If refrigeration temperature is lower than -10 °C, problem of mass transfer will happen because of the low evaporation temperature in sorption process.

For the composite sorbent with activated carbon as the matrix, its performance is also a combination of physical adsorption and chemisorption. Figure 3.32 shows the isobaric sorption performance of chemical sorbent ($CaCl_2$) and composite sorbent (composite of $CaCl_2$ and activated carbon) under the same quality of chemical sorbent and the same filling amount. The physical sorbent inside the composite sample 1 is 14–28 mesh activated carbon produced from Hainan coconut shell, and the volume ratio of calcium chloride to activated carbon is 2:1 (mass ratio 4:1). Adequately ensure the sorption time at different temperatures to eliminate the influence of heat and mass transfer on the isobaric sorption performance. If the dynamic characteristics of the composite sorbent are assumed to be a simple combination of chemisorption and physical adsorption, the sorption capacity should be the sum of the chemisorption

Table 3.14 Performance of simply mixed composite sorbent for ice making condition (compared with pure $CaCl_2$) [52]

	$CaCl_2$ ratio (%)	Activated carbon ratio (%)	Increment of filling volume (%)	Increment of volume cooling capacity (%)	Refrigeration temperature (°C)
Condition 1	60	40	29	29	-15
Condition 2	70	30	40	40	-15
Condition 3	80	20	51	38	-15

Table 3.15 Performance of consolidated composite sorbent for ice making condition (compared with pure $CaCl_2$) [52]

	$CaCl_2$ ratio (%)	Activated carbon ratio (%)	Water ratio (%)	Cement ratio (%)	Increment of filling volume (%)	Increment of volume cooling capacity (%)	Refrigeration temperature (°C)
Condition 1	49	32.7	12.2	6.1	29	29	-15
Condition 2	55.4	23.8	13.9	6.9	40	40	-10
Condition 3	52.2	28.3	13	6.5	34	34	-15

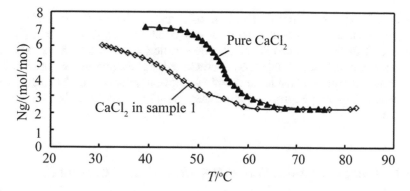

Fig. 3.32 Isobaric sorption performance of sample 1 and CaCl₂ [52]

capacity and the physical adsorption capacity, and the chemisorption curve of $CaCl_2$ inside the composite sorbent should be the same or similar. However, Fig. 3.32 shows that the sorption curve of pure $CaCl_2$ at 55 °C is close to the abscissa of the vertical line. Therefore, this is a chemisorption process controlled by a single variable. But for the sorption of composite materials, due to the addition of physical sorbents, it is controlled by two variables, showing obvious characteristics. This means that its sorption performance is affected by the restraint pressure and sorption temperature.

The deviation of chemisorption process of $CaCl_2$ inside composite sorbent is mainly caused by the physical sorbent in the composite sorbent. In the sorption process of the composite sorbent, the combination of the capillary condensation process of refrigerant in the physical sorbent and the chemical sorption process makes the chemical sorption precursor states deviate from the chemisorption theory. In order to have a clear understanding for that, the composite sample 2 is produced by 20–40 mesh activated carbon and $CaCl_2$. The mass ratio and filling amount of the additives in the bed are the same as in sample 1. Figure 3.33 shows the kinetic results of the two samples at 430 kPa evaporation pressure.

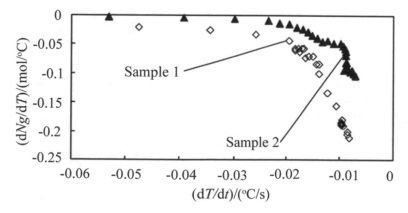

Fig. 3.33 Performance comparison of sample 1 and sample 2

Figure 3.33 presents that different composite sorbents have different performance. The performance of the composite sample 1 is better than that of the composite sample 2. This indicates that the composite sorption kinetics should combine with chemical sorption kinetics and mass transfer kinetics of the porous medium. But how to separate the sorption kinetics of chemical sorption process with the mass transfer process and sorption performance of different granular activated carbon is a difficult problem.

3.3.2 Composite Solid Sorbents with Activated Carbon Fiber

If activated carbon fiber is selected as the matrix of the composite sorbent, a sample with high thermal conductivity is preferred, which will effectively improve the heat transfer of the chemical sorbent. Generally, the higher the graphitization degree of activated carbon fiber, the better the thermal conductivity. However, because the price of activated carbon fiber is very expensive, the best activated carbon fiber should always have a relatively low price and a high degree of graphitization.

Activated carbon fiber accounts for 25% of ICF and GFIC respectively. In IFC, if salt does not fall off the activated carbon fiber, and if ammonia is used as a refrigerant, the sorption performance and cooling process of ICF and GFIC are shown in Figs. 3.34 and 3.35. The performance of the two working pairs is similar.

Making the comparison between the performance of ICF and GFIC, the results are shown in Table 3.16.

Fig. 3.34 Sorption performance of ICF and GFIC [53, 54]

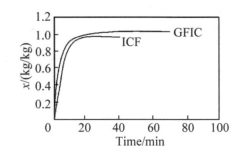

Fig. 3.35 Temperature evolution of ICF and GFIC for the refrigeration process [53, 54]

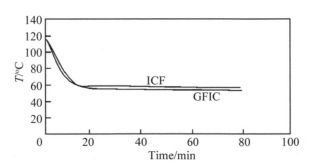

Table 3.16   Advantages and disadvantages between GFIC and ICF [53]

Composites	Advantages	Disadvantages
GFIC	Good dynamic performance and sorption performance Salt particles evenly distribute in the activated carbon fiber	Preparing time is long Intercalation salt is hard to control Requires high degree of graphitization of the activated carbon fiber
IFC	Good dynamic performance and sorption performance Easier preparing process Easy to control the proportion of salts in the activated carbon Doesn't require high degree of graphitization of the activated carbon fiber	Sorption performance slightly lower than GFIC Salt particles is easy to fall down from the activated carbon fiber

3.4   Properties of Composite Solid Sorbents with Silica Gel

3.4.1   Composite Sorbents of Silica Gel and CaCl₂

The composite sorbent based on silica gel is usually prepared by the impregnation method [55–58]. The silica gel is immersed in a halide solution with fixed solubility (such as $CaCl_2$), and then the silica gel is dried to obtain a composite sorbent with good sorption capacity. In fact, the composite sorbent based on silica gel is to immerse the halide chemical sorbent into the micropores of silica gel to obtain a strong sorption capacity. Sorption and desorption reactions may occur between the composite sorbent and water. In actual use, it is required that the halide solution after water sorption does not flow out of the silica gel pores, for which special processing techniques are required.

Figure 3.36 shows the sorption performance of different proportions of salt on the composite sorbent under environmental conditions, where S is the sorbent sample, and the number after S is the concentration of the calcium chloride solution used to prepare the composite sorbent. For example, S0 is a silica gel matrix, and S40 is a

Fig. 3.36   Sorption isothermal curves of different samples [59, 60]

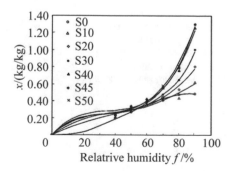

composite sorbent with a concentration of 40% $CaCl_2$. It can be seen from Fig. 3.36 that when the halide concentration is greater than 40%, the equilibrium sorption capacity does not increase much. The higher the concentration of $CaCl_2$, the easier it is to liquefy. Therefore, the best composite sorbent is sample S40.

For the composite sorbent with the silica gel as matrix (Aristov defined it as selective water sorbent SWS), the chemical sorbent is impregnated in the micropores of the porous medium which has sorption reaction, showing the characteristics between a porous silica gel and pure hygroscopic halide [55, 61].

Figure 3.37 shows the sorption properties of the composite sorbent with the silica gel of average pore radius of 7.5 nm and $CaCl_2$ proportion of 33.7% in SWS. Figure 3.37 shows that the SWS sorption performance has been improved a lot if compared with silica gel. The maximum sorption capacity of SWS and silica gel are 0.75 and 0.1 kg/kg, respectively. In addition, it also shows that for the low sorption capacity, the chemical sorption phenomenon is different from physical adsorption. When the sorption capacity is 0.11 kg/kg, there is a sorption platform, which indicates that the sorption capacity during this process is only associated with pressure but has nothing to do with the sorption temperature, i.e., it is controlled by single parameter. While the sorption capacity is higher than 0.11 kg/kg, the sorption process is controlled by the dual variables of sorption temperature and constraint pressure.

The sorption platform in Fig. 3.37 states that $CaCl_2$ embedded in porous silica gel does not affect the sorption properties, but differs from pure $CaCl_2$. The sorption characteristics of solid $CaCl_2$ crystal hydrate change because halides impregnate inside the micro pores of silica gel. SWS can form bi-hydrates and tri-hydrates with very low vapor pressure. The vapor pressure in the hydrate formation process of SWS is an order of magnitude lower than that of the $CaCl_2$ hydrate formation process.

Using other chemical sorbents such as LiBr, LiCl etc., it will show the common characteristics similar with $CaCl_2$ for the performance of composite sorbent. The monohydrate forms during the phase of small sorption capacity. When the sorption is larger, compared with the phase for the small sorption capacity, the characteristics of monohydrate will change significantly, which is mainly caused by the impregnating process of the inorganic halides in the porous matrix. The impregnation process of the halide in the porous medium won't influence the sorption performance of halide very much.

Fig. 3.37 Comparison of sorption performance between SWS and silica gel under the condition of 25 mbar vapor pressure [55]

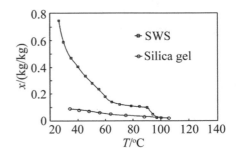

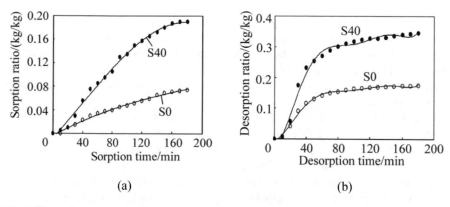

Fig. 3.38 Performance comparison of composite sorbents and silica gel [60]. **a** Sorption process under the conditions of 10 °C freezing water and 30 °C cooling water, **b** Desorption process under the conditions of 80 °C heat source

In general, the sorption and desorption dynamics of composite sorbents are better than the pure silica gel. Taking S40 composite sorbent by Daou from SJTU for example [60], its kinetic characteristics are compared with pure silica gel, and the results are shown in Fig. 3.38. It could be seen that the sorption and desorption rate of composite sorbent is much faster than that of the pure silica gel.

3.4.2   The Composite Sorbent of Silica Gel and LiCl

The mass concentration of the impregnation solution and the pore structure of the matrix are two main parameters that will have a strong influence on the properties of the prepared composite sorbent. As listed in Table 3.17, two types of mesoporous silica gel (types A and C) and four concentrations of LiCl solutions (10, 20, 30 and 40 wt.%) are obtained. They are used in manufacturing 8 samples. For both silica gel type A and type C, the LiCl mass concentration in the complex increases with the LiCl concentration in the impregnation solution, as expected. Due to the larger pore size and pore volume, silica gel type C has an internal space, especially 30 and 40 wt.%. More LiCl crystals can be carried at high concentration. The composite

Table 3.17 LiCl mass concentration in the composite sorbents prepared by different silica gel pore sizes and impregnating LiCl mass concentrations

Impregnating LiCl mass concentration in solution (%)		10	20	30	40
LiCl mass concentration in composite sorbent	Silica gel type A (pore size 2–3 nm) (%)	6.5	19.8	24.3	25.6
	Silica gel type C (pore size 8–12 nm) (%)	11.4	24.0	35.1	43.6

sorbent developed in silica gel type C is selected as the study sample because the higher halide content in silica gel causes a higher water sorption potential. The four compounds with four concentrations are abbreviated as SLi10, SLi20, SLi30 and SLi40, and the last two numbers represent the impregnated LiCl mass concentration. Pure silica gel type C is called SG and its properties are also studied for comparison with composites.

The water sorption isobars of SG and four kinds of silica gel-LiCl composite materials at 0.88, 1.66, 2.88 and 4.45 kPa are shown in Fig. 3.39. In Fig. 3.39a, the isobar of SG is very smooth. Water sorption is a function of temperature and pressure. Unfortunately, this mesoporous silica gel exhibits poor water sorption performance under experimental conditions, with a maximum water sorption of 0.088 g/g at 30 °C and 2.88 kPa. Facts have shown that this type of silica gel is not sufficient for water sorption alone. The isobars of the composite materials in Fig. 3.39b–e show a similar trend: a sudden change in the slope may be found at high temperature, which indicates that some transition has occurred in this range. After the transition zone, water sorption will drop to almost zero. However, no steady state has been observed indicating the formation of halide hydrates, which have been found in composite materials that have been immersed in other halides such as $CaCl_2$ [32, 33]. It can be understood that as the halide content in the composite increases, the water sorption will also increase. For SLi10, the maximum water sorption is 0.467 g/g, which is lower than its pore volume 0.91 cm^3/g, so there is no need to worry about residual problems.

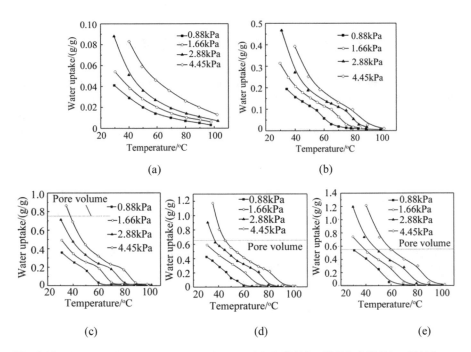

Fig. 3.39 Water sorption isobars on the sorbents: **a** SG; **b** SLi10; **c** SLi20; **d** SLi30; **e** SLi40

Under the conditions of 40 °C and 4.45 kPa, the maximum water sorption of SLi40 exceeds 1.2 g/g, but it should be noted that since the pore volume of SLi40 is only 0.55 cm^3/g, it will encounter residual problems. SLi20 and SLi30 will also encounter the same problems as SLi40. When it comes to specific working conditions (such as a closed sorption system), these sorbents can be selected to find the most suitable sorbent. If the maximum water sorption occurs at the lowest temperature of 30 °C and the highest pressure of 1.88 kPa (evaporating temperature of 15 °C), it can be concluded that SLi30 is the best choice because it has the largest water sorption and there is no need to worry about carryover. Another advantage of composite sorbents is that they can achieve complete water desorption at relatively low temperatures of 60–100 °C, which means they can be regenerated by conventional low-temperature heat sources.

3.4.3 The Comparison Between Activated Carbon Fiber and Silica Gel as Matrix

3.4.3.1 Comparison of Different Isothermal Sorption Curves

Micromeritics ASAP2020 gas sorption analyzer is adopted to measure specific surface area using the conventional BET method and porous characteristics of the composites using the BJH (Barret-Joyner-Halenda) method by the standard nitrogen sorption/desorption measurement [62]. The sorption isotherms are implemented at 77 K which is the temperature of liquid nitrogen. The sorption isotherms of pure ACF and silica gel exhibit exactly two kinds of curves (Fig. 3.40a, b), and it is mainly related to the difference of pore structure and size between silica gel and ACF. ACF belongs to micropore medium with pore diameter of 1.58 nm, which pertains

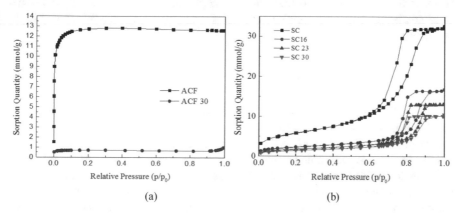

(a)   (b)

Fig. 3.40 Isotherms by nitrogen sorption at 77 K. **a** Sorption quantity of ACF and ACF 30, **b** Sorption quantity of SC, SC16, SC23 and SC30 [62]

to the same order of magnitude as the nitrogen molecule 0.304 nm. So the potential energy fields overlap between the formation of a strong energy field to attract other molecules, which enhances the interaction between the solid surface and the molecules. Therefore, the sorption isotherms of ACF rise sharply from the beginning, and become saturated later because molecules have filled pore channels. Silica gel belongs to the mesoporous materials with the average pore diameter of 7.14 nm, which is much greater than the diameter of nitrogen molecules, so it presents a typical multi-molecular sorption.

Figure 3.40a also shows that the specific surface area of ACF is significantly reduced after it is impregnated with calcium chloride. The specific surface area of silica gel declines with the increasing mass proportion of calcium chloride, and the certain proportional relationship can be seen in Fig. 3.40b.

3.4.3.2   Comparison of Impregnation of CaCl$_2$ Measured by ICP

Three groups are set to research the effect of the fiber direction on salt impregnation, and the ratios of halide are determined by:

$$R_i = (M_{Ci} - M_{Mi})/M_{Ci} \tag{3.17}$$

where R_i is the increase ratio of calcium chloride, M_{ci} is the mass of dry composite sample (kg), M_{Mi} is the weight of dry host matrix (silica or ACF).

Generally, the mass of CaCl$_2$ remaining in the ACF felts and silica pores could be calculated by measuring the difference in dry samples mass before and after the impregnation. However, such a method isn't quite accurate because the CaCl$_2$ will sorb water vapor in air when it exposed to the atmosphere. Meanwhile, CaCl$_2$ inevitably has the crystal water, which only can be dried out when the temperature is above 300 °C. In order to study the impregnation of CaCl$_2$ in detail, inductive coupled plasma emission spectrometer (ICP) is used to measure the calcium chloride accurately. The equation is:

$$M_{CaCl_2} = M_{Ca} \times \frac{CaCl_2(mol)}{Ca(mol)} \tag{3.18}$$

In Fig. 3.41, ACF-A and ACF-B are the same material ACF. CaCl$_2$ content of ACF-A is determined by electronic weighing scales, while CaCl$_2$ content of ACF-B is determined by ICP. Similarly, SC-A and SC-B are the same material SC, and CaCl$_2$ content of SC-A is determined by electronic weighing scales, while CaCl$_2$ content of SC-B is determined by ICP. It shows that results of method A (electronic weighing scales) are higher than method B (ICP), which means the dry calcium chloride of samples contains some crystallization water. Method B is a direct measurement of calcium, thus it is more accurate than method A.

Fig. 3.41 Comparison of results between weighing method and ICP methods [62]

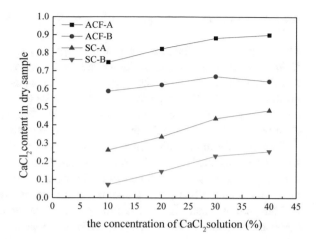

SC-B shows that the $CaCl_2$ content of SC is closely related to the concentration of solution, and it increases with the increase of the concentration of $CaCl_2$ solution, but the growth rate decreases at high concentration. ACF-B shows that $CaCl_2$ content has little change when solution concentration changes. For example, when the concentration is 0.4, $CaCl_2$ content is 0.61, which is a little bit less than the value of 0.67 for the $CaCl_2$ concentration of 0.3. Compared with SC-B, $CaCl_2$ content of ACF-B are 6 times higher than SC-B at 0.1 concentration of $CaCl_2$ solution, and two times higher at 0.4 concentration of $CaCl_2$ solution. In terms of salt impregnation, performance of ACF is better than SC.

Actually, the impregnated salt is in crystalline hydrate form existing in matrix (ACF or SC). The difference between ACF-A and ACF-B is the amount of water of crystallization, $CaCl_2 \cdot nH_2O$, n can be obtained by

$$n = \frac{m_{ACF/SC-A} - m_{ACF/SC-B}}{CaCl_2(mol) \times M_{H_2O}} \qquad (3.19)$$

Table 3.18 shows the number of crystallization water, and the values for the ACF is generally lower than that for SC. It should be benefit from the microstructure and heat transfer capability of ACF. Another advantage is that ACF materials can be heated at high temperature for a long time and not be burned, so ACF is fit for wider temperature range and is provided with larger water desorption potential.

Table 3.18 Water of crystallization of the impregnated halide [62]

Concentration	10%	20%	30%	40%
Crystallizing water in ACF composite sorbent	1.69	1.98	1.97	2.91
Crystallizing water in SC composite sorbent	10.43	8.15	5.53	5.46

3.4.3.3   Comparison of Non-equilibrium Sorption Performance

The eight kinds of sorbents are tested in the constant temperature and humidity chamber. Electronic balance scales are adopted to measure weight of all the samples at different time point. And the weighing interval ranges gradually from 15, 20, 30, 60 to 80 min. Sorption equilibrium is defined when the measured weights of two consecutive interval of 80 min are less than 3%. The temperature accuracy of the chamber is ± 0.5 °C while the relative humidity deviation is $\pm 3\%$RH.

Non-equilibrium sorption properties are shown in Fig. 3.42. The ordinate represents the ratio of the increment of the sample in the different processes of experiments and the initial weight is obtained from the sample dried under 120 °C for 4 h. SA means the type A silica gel with 3–5 mm grain size and the average pore diameter of 2–3 nm, and SC means the type C silica gel. Both of them are set as the experimental comparison groups without $CaCl_2$. It can be concluded that the higher the concentration of calcium chloride solution, the more water uptake by per unit mass sorbents. Sorption performance of SC30 is 6–7 times higher than the pure type C silica gel,

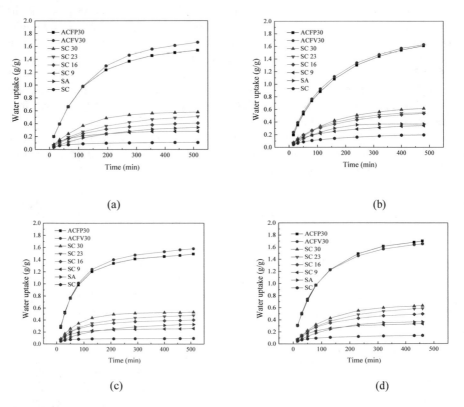

Fig. 3.42 Non-equilibrium sorption performance of the composites under the condition of sorption temperatures. Pure SA and SC under the sorption condition of 20–35 °C temperature and 70% relative humidity; **a** 20 °C, **b** 25 °C, **c** 30 °C, **d** 35 °C [62]

and is two or three times higher than the pure type A silica gel. Due to the notable increment of calcium chloride, the performance of ACF30 is three times better than the performance of SC30, even though both of them are impregnated with the same concentration of $CaCl_2$.

It is worth mentioned that SC30 composite sorbents appear the different levels of carryover problem after 200 min under the experimental conditions. After several cycles of sorption and desorption, parts of silica gel composite sorbents rupture, and the other parts look complete but have cracks on the surface.

References

1. Krzesinska M, Celzard A, Mareche JF, Puricelli S (2001) Elastic properties of anisotropic monolithic samples of compressed expanded graphite studied with ultrasounds. J Mater Res 16(2):606–614
2. Celzard A, Mareche JF, Furdin G, Puricelli S (2000) Electrical conductivity of anisotropic expanded graphite-based monoliths. J Phys D Appl Phys 33(23):3094–3101
3. Celzard A, Krzesinska M, Mareche JF, Puricelli S (2001) Scalar and vectorial percolation in compressed expanded graphite. Phys A 294:283–294
4. Krzesinska M (2004) Influence of the raw material on the pore structure and elastic properties of compressed expanded graphite blocks. Mater Chem Phys 87:336–344
5. Krzesinska M, Lachowski AI (2004) Elastic properties of monolithic porous blocks of compressed expanded graphite related to their specific surface area and pore diameter. Mater Chem Phys 86:105–111
6. Wang LW, Tamainot-Telto Z, Metcalf SJ, Critoph RE, Wang RZ (2010) Anisotropic thermal conductivity and permeability of compacted expanded natural graphite. Appl Therm Eng 30(13):1805–1811
7. Tian B, Jin ZQ, Wei DS, Wang LW, Wang RZ (2010) Testing on permeability of compacted graphite (in Chinese). CIESC J 21:35–38
8. Han JH, Cho KW, Lee KH, Kim H (1998) Porous graphite matrix for chemical heat pumps. Carbon 36(12):1801–1810
9. Bonnissel M, Luo L, Tondeur D (2001) Compacted exfoliated natural graphite as heat conduction medium. Carbon 39:2151–2161
10. Wang LW, Metcalf SJ, Thorpe R, Critoph RE Tamainot-Telto Z (2011) Thermal conductivity and permeability of consolidated expanded natural graphite treated with sulphuric acid. Carbon 49(14):4812–4819
11. Tamainot-Telto Z, Critoph RE (2001) Monolithic carbon for sorption refrigeration and heat pump applications. Appl Therm Eng 21:37–52
12. Takahashi Y, Azumi T, Sekine Y (1989) Heat capacity of aluminum from 80 to 880 K. Thermochim Acta 139:133–137
13. Jin ZQ, Tian B, Wang LW, Wang RZ (2013) Comparison on thermal conductivity and permeability of granular and consolidated activated carbon for refrigeration. Chin J Chem Eng 21(6):676–682
14. Jin JQ, Wang LW, Jiang L, Wang RZ (2013) Experiment on the thermal conductivity and permeability of physical and chemical compound adsorbents for sorption process. Heat Mass Transfer 49:1117–1124
15. Biloe S, Goetz V, Mauran S (2001) Characterization of adsorbent composite blocks for methane storage. Carbon 39:1653–1662
16. Wang LW, Metcalf SJ, Thorpe R, Critoph RE, Tamainot-Telto Z (2012) Development of thermal conductive consolidated activated carbon for adsorption refrigeration. Carbon 50:977–986

17. Tamainot-Telto Z, Critoph RE (1997) Adsorption refrigerator using monolithic carbon-ammonia pair. Int J Refrig 20(2):146–155
18. Critoph RE (1988) Performance limitations of adsorption cycles for solar cooling. Sol Energy 14(1):21–31
19. Eun TH, Song HK, Han JH, Lee KH, Kim JN (2000) Enhancement of heat and mass transfer in silica-expanded graphite composite blocks for adsorption heat pumps: part I. Characterization of the composite blocks. Int J Refrig 23:64–73
20. Eun TH, Song HK, Han JH, Lee KH, Kim JN (2000) Enhancement of heat and mass transfer in silica-expanded graphite composite blocks for adsorption heat pumps. Part II. Cooling system using the composite blocks. Int J Refrig 23:74–81
21. Zheng X, Wang LW, Wang RZ, Ge TS, Ishugah TF (2014) Thermal conductivity, pore structure and adsorption performance of compact composite silica gel. Int J Heat Mass Transf 68:435–443
22. Shim JH, Park M, Lee YH, Kim S, Im YH, Suh JH, Cho YW (2014) Effective thermal conductivity of MgH$_2$ compacts containing expanded natural graphite under a hydrogen atmosphere. Int J Hydrogen Energy 39:349–355
23. Wang K, Wu JY, Wang RZ, Wang LW (2006) Composite adsorbent of CaCl$_2$ and expanded graphite for adsorption ice maker on fishing boats. Int J Refrig 29:199–210
24. Wang K, Wu JY, Wang RZ, Wang LW (2006) Effective thermal conductivity of expanded graphite-CaCl$_2$ composite adsorbent for chemical adsorption chillers. Energy Convers Manage 47:1902–1912
25. Oliveira RG, Wang RZ (2007) A consolidated calcium chloride-expanded graphite compound for use in sorption refrigeration systems. Carbon 45:390–396
26. Zajaczkowski B, Królicki Z, Jezowski A (2010) New type of sorption composite for chemical heat pump and refrigeration systems. Appl Therm Eng 30:1455–1460
27. Huang HJ, Wu GB, Yang J, Dai YC, Yuan WK, Lu HB (2004) Modeling of gas solid chemisorption in chemical heat pumps. Sep Purif Technol 34:191–200
28. Mazet N, Meyer P, Neveu P, Spinner B (1994) Concept and study of a double effect refrigeration machine based on the sorption of solid and ammonia gas and controlled by heat pipes. In: International absorption heat pump conference. ASME, New Orleans, LA, USA, pp 407–412
29. Hosatte S, Rheault F (1992) Kinetics and modelling of CaCl$_2$-NH$_3$ reactions. In: Proceeding of the symposium of solid sorption refrigeration, vol 1. LIMSI, Paris, pp 245–252
30. Valkov V (1993) Measurement of technique for determination of thermochemical properties of thermochemical storage materials, experimental heat transfer. Fluid Mech Thermodyn 12(6):529–536
31. Fujioka K, Suzuki H (2013) Thermophysical properties and reaction rate of composite reactant of calcium chloride and expanded graphite. Appl Therm Eng 50:1627–1632
32. Jiang L, Wang LW, Jin ZQ, Wang RZ, Dai YJ (2013) Effective thermal conductivity and permeability of compact compound ammoniated salts in the adsorption/desorption process. Int J Therm Sci 71:103–110
33. Mauran S, Prades P, Haridon FL (1993) Heat and mass transfer in consolidated reaction beds for thermochemical systems. Heat Reco Syst CHP 13:315–319
34. Mazet N, Amouroux M (1991) Analysis of heat transfer in a non-isothermal solid-gas reacting medium. Chem Eng Commun 99:175–200
35. Mazet N, Lu HB (1998) Improving the performance of the reactor under unfavorable operating conditions of low pressure. Appl Therm Eng 18(9–10):819–835
36. Han JH, Lee KH (2001) Gas permeability of expanded graphite-metallic salt composite. Appl Therm Eng 21:453–463
37. Lu HB, Mazet N, Spinner B (1996) Modelling of gas-solid reaction-coupling of heat and mass transfer with chemical reaction. Chem Eng Sci 51:3829–3845
38. Han JH, Lee KH, Kim H (1999) Effective thermal conductivity of graphite-metallic salt complex for chemical heat pumps. J Thermophys Heat Transfer 13(4):481–488
39. Bou P, Moreau M, Prades P (1999) Active composite with foliated structure and its use as reaction medium. United States Patent 5861207

40. Jiang L, Wang LW, Wang RZ (2014) Investigation on thermal conductive consolidated composite $CaCl_2$ for adsorption refrigeration. Int J Thermal Sci 81:68–75
41. Jiang L, Wang LW, Jin ZQ, Tian B, Wang RZ (2012) Permeability and thermal conductivity of compact adsorbent of salts for sorption refrigeration. ASME—J Heat Transfer 134:104503–104506
42. Zhou ZS, Wang LW, Jiang L et al (2016) Non-equilibrium sorption performances for composite sorbents of chlorides—ammonia working pairs for refrigeration. Int J Refrig 65:60–68
43. Gao J, Wang LW, Wang RZ et al (2017) Solution to the sorption hysteresis by novel compact composite multi-salt sorbents. Appl Therm Eng 111:580–585
44. Goetz V, Marty A (1992) A model for reversible solid-gas reactions submitted to temperature and pressure constraints: simulation of the rate of reaction in solid-gas reactor used as chemical heat pump. Chem Eng Sci 47:4445–4454
45. Wang LW, Wang RZ, Wu JY, Wang K (2005) Studies on chemical adsorption hysteresis for adsorption refrigeration. J Eng Thermophys-Rus 6:901–904
46. Trudel J, Hosatte S, Ternan M (1999) Solid-gas equilibrium in chemical heat pumps: the NH_3-$CoCl_2$ system. Appl Therm Eng 19:495–511
47. Aidoun Z, Ternan M (2002) The synthesis reaction in a chemical heat pump reactor filled with chloride salt impregnated carbon fibres: the NH_3-$CoCl_2$ system. Appl Therm Eng 22:1943–1954
48. An GL, Wang LW, Gao J, Wang RZ (2019) Mechanism of hysteresis for composite multi-halide and its superior performance for low grade energy recovery. Scienti Rep 9(1):1563
49. Wang LW, Wang RZ, Wu JY, Wang K (2004) Compound adsorbent for adsorptin ice maker on fishing boats. Int J Refrig 27(4):401–408
50. Wang LW, Wang RZ, Wu JY, Wang K (2004) Adsorption performances and refrigeration application of adsorption working pair of $CaCl_2$-NH_3. Sci China Ser E 47(2):173–185
51. Wang LW, Wang RZ, Wu JY, Wang K (2005) Research on the chemical adsorption precursor state of $CaCl_2$-NH_3 for adsorption refrigeration. Sci China Ser E 48(1):70–82
52. Wang LW (2005) Performances, mechanisms, and application of a new type compound adsorbent for efficient heat pipe type refrigeration driven by waste heat (in Chinese, PhD thesis). Shanghai Jiao Tong University, Shanghai, China
53. Dellero T, Sarmeo D, Touzain P (1999) A chemical heat pump using carbon fibers as additive. Part I: enhancement of thermal conduction. Appl Therm Eng 19:991–1000
54. Dellero T, Touzain P (1999) A chemical heat pump using carbon fibers as additive. Part II: study of constraint parameters. Appl Therm Eng 19:1001–1011
55. Aristov YI, Restuccia G, Caccioba G et al (2002) A family of new working materials for solid sorption air conditioning systems. Appl Therm Eng 22:191–204
56. Tokarev M, Gordeeva L, Romannikov V, Glaznev I, Aristov YI (2002) New composite sorbent $CaCl_2$ in mesopores for sorption cooling/heating. Int J Thermal Sci 41:470–474
57. Levitskij EA, Aristov YI, Tokarev MM et al (1996) Chemical heat accumulators: a new approach to accumulating low potential heat. Solar Energy Solar Cells 44:219–235
58. Restuccia G, Freni A, Vasta S, Aristov YI (2004) Selective water sorbent for solid sorption chiller: experimental results and modeling. Int J Refrig 27:284–293
59. Vasiliev LL, Mishkinis DA, Antuh A, Snelson K, Vasiliev Jr LL (1999) Multisalt-carbon chemical cooler for space applications. In: Proceedings of international absorption heat pump conference, Munich, Germany, pp 579–583
60. Daou K (2006) The development, experiment, and simulation of a new type of efficient composite adsorbent driven by the low grade heat source (in Chinese, PhD thesis). Shanghai Jiao Tong University, Shanghai, China
61. Aristov YI, Tokarev MM, Parmon VN, Restuccia G, Burger HD et al (1999) New working materials for sorption cooling/heating driven by low temperature heat: properties. In: Proceedings of international sorption heat pump conference, Munich, Germany, pp 24–26
62. Wang JY, Wang RZ, Wang LW (2016) Water vapor sorption performance of ACF-$CaCl_2$, and silica gel-$CaCl_2$, composite adsorbents. Appl Therm Eng 100:893–901

Chapter 4
Kinetics of Solid Composite Sorbents

Abstract In this chapter, the typical equilibrium principles and phenomena in halide-ammonia based composite sorbents are introduced. Advantages and shortcomings of several classical non-equilibrium models are analyzed. After that several recently discovered sorption phenomena are discussed, and finally, several new directions for the development of kinetic models for halide-ammonia based composite sorbents are proposed.

Keywords Complexation mechanism · Principle · Precursor state · Hysteresis · Halide · Ammonia · Kinetic model · Analogical model · A phenomenological model

Solid sorption can be divided into physical adsorption [1–3] and solid chemisorption [4–6]. Physical adsorption is a consequence of the van der Waals forces [7], while solid chemisorption is a complexation reaction between the surface molecules of sorbents and sorbates [8]. Compared with physical adsorption, solid chemisorption has a larger sorption capacity, which is beneficial for increasing the specific cooling power (SCP) and decreasing the volume of the sorption reactor.

Solid chemisorption working pairs consist of oxide-oxygen, hydride-hydrogen, and metal halide-ammonia. Hydride-hydrogen is mainly used for the storage of hydrogen, and metal halide-ammonia working pairs are extensively used for heat pump, refrigeration and thermal energy storage [9]. To improve the heat and mass transfer performance of sorbents, as analyzed in the previous chapter, composite sorbent matrices have been widely studied [10–16] in the past three decades.

The classical solid sorption theories include the Langmuir sorption theory [17], Gibbs approach [18], Polanyi's potential theory [19–21], and Brunauer–Emmett–Teller (BET) equation [22]. Among these theories, the BET method is used to research the surface area of porous materials [23]. The Dubinin-Radushkevich (D-R) equation [24, 25], established based on the Polanyi potential theory, is applied in physical adsorption. For materials with novel structures, a physical model 'multilayer model with saturation' has been successfully used to interpret the equilibrium sorption of ethanol on metal–organic frameworks [26]. For composite solid sorbents, it was

© Science Press 2021
L. Wang et al., *Property and Energy Conversion Technology of Solid Composite Sorbents*, Engineering Materials,
https://doi.org/10.1007/978-981-33-6088-4_4

recently found similar to physical adsorption that kinetics can show the existence of multiple pores in the matrix [27, 28], and the D-R (Dubinin-Astakhov (D-A)) equation is also used for establishing kinetic models [29].

Since the 1990s, several types of kinetic models have been established to explain the solid chemisorption mechanisms. Crozat and Stitou divided solid chemisorption kinetic models into three groups, i.e., local models, global models, and analytical models [29]. Local models [30–32] are based on the local heat and mass transfer and given uniform variables in small volumes; the problems in time and space need to be numerically solved since they can generate several partial derivatives. Global models [33–37] consider uniform values on the scale of the reactor. Since global transformations of heat conductivity, permeability, and thermal capacity are averaged in the volume of the reactor, these models lead to a set of differential equations, which only need discretization in the time scale of the problems. Analytical models [29, 38] consider the variables to be averaged during the entire reaction time, which means that the variables are only relevant to the space variables. In some cases, by considering that mass transfer in the reaction medium will not impose any limitations, only the heat transfer equation needs to be solved, and an analytical resolution can be obtained (the kinetic equation for the reaction process will not be considered since no time variables are included in this kind of model).

On the other hand, according to Spinner and Lebrun's analysis [39], the kinetic model should be either analogical or phenomenological. The analogical approach, which does not involve detailed mechanisms, is only used to reproduce the effect of solid sorbents and overall performance. Since the reaction medium is considered a uniform entity, the kinetic models are no longer based on the structure and properties of the medium, but on homogeneous reactions and physicochemical processes, which consist of several parameters. As the parameters are not evolved from the properties of the medium, they are experienced parameters and must be identified with the experimental values. Therefore, these models are not knowledge models. The phenomenological models link with the precision phenomena and physicochemical properties, so they are counted as knowledge models. On the other hand, since the structure and properties of the reaction medium need to be known in detail, it makes the model too tricky and complex to manipulate. Under this criterion, phenomenological models [32, 40] are much less than the analogical models.

In the last 20 years, many reviews on sorption refrigeration have been published. Most of the reviews published in recent years concentrating on the solar sorption refrigeration [41–45], especially on the physical adsorption equilibrium models and principles used in solar sorption refrigeration systems [46]. What is more, several kinetic models used for various chemisorption working pairs have been applied in recent years. Jiang and Guo [47] used the Sudden Vector Projection (SVP) model to predict bond selectivity, mode specificity, surface lattice effects and normal scaling behavior in the case of water dissociative chemisorption on Cu(111), Ni(111), Pt(111), and Pt(110)-(1 × 2) using direct plane-wave density functional theory. Later, they investigated site-specific reaction probabilities on a nine-dimensional global potential energy surface (PES) using a quasi-seven-dimensional quantum dynamic model [48]. It is shown that the site-specific reactivity is controlled mainly by the topography of the PES instead of the barrier height. Similarly, Jackson et al. [49] worked on the dissociative chemisorption of methane on metal surfaces, and Liu et al. [50] studied H_2 sorption on the $Cu(111)$ surface. In both cases, the quantum model was used.

Even though halide-ammonia working pairs are widely applied in refrigeration, heat pump, and thermal energy storage, there have been very few studies analyzing the chemisorption kinetic models of halide-ammonia working pairs in detail. However, such studies are essential for designing the reactors and for experimental analysis. In order to help other researchers to achieve a clear understanding of the mechanism of halide-ammonia working pairs, in this chapter, we initially introduce the typical equilibrium principles and phenomena in halide-ammonia working pairs. Next, the advantages and shortcomings of several classical non-equilibrium models are analyzed. Later, several recently discovered phenomena are discussed, and at last, we propose some directions for the development of kinetic models for halide-ammonia working pairs.

4.1 Typical Principles and Phenomena

4.1.1 Clapeyron Equation

The reaction between the halide and ammonia can be written as

$$M_aCl_b(NH_3)_n + (m - n)NH_3 \Leftrightarrow M_aCl_b(NH_3)_m + n\Delta H_r \qquad (4.1)$$

where M represents the metal, and ΔH_r is the reaction enthalpy (J/mol). The values of m and n can be determined by the properties of different halides. Through detailed derivation, the Clapeyron equation for halide-ammonia working pairs has been proposed

$$\ln p_{NH_3} = -\frac{\Delta H^0}{RT} + \frac{\Delta S^0}{R} \qquad (4.2)$$

where p_{NH_3} is the average pressure in evaporator/condenser (101,325 Pa), ΔH^0 and ΔS^0 are the changes in the standard enthalpy and entropy, respectively, R is the universal gas constant, and T represents the reaction temperature in the reactor (298.15 K).

This equation is valid under average temperature and pressure, while the thermal capacity should be considered under other conditions. Hutting and Biltz proposed an experimental equation (Eq. 4.3) as follows.

Fig. 4.1 The equilibrium
reaction lines of halides [52]

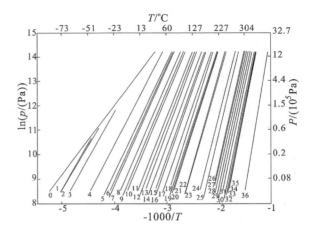

$$\ln p_{NH_3} = -\frac{\Delta H^0}{RT} + 1.75 \ln T + aT + 3.3 \ln 10 \qquad (4.3)$$

Depending on the type of working pairs, the parameter 'a' can vary between
−0.0024 and −0.0017.

Touzain [51] reviewed over 350 kinds of halide-ammonia working pairs and found
that the mean difference of enthalpy was −130 kJ/mol. The modified equation to
calculate the reaction enthalpy is shown in Eq. 4.4.

$$\Delta H_{cal}^0 = \Delta H_{exp}^0 - T_{exp}(\Delta S_{exp}^0 + 130) \qquad (4.4)$$

However, to simplify the calculation process, most researchers still use Eq. 4.2 to
analyze experimental data.

The equilibrium pressures for various halide-ammonia working pairs are shown
in Fig. 4.1 [52] and the reaction enthalpy, entropy, and specific heat capacity are
listed in Table 4.1 [52].

The sorbents with reasonable refrigeration performance are listed in Table 4.2
[51] together with the energy per kilogram sorbent. Table 4.2 also lists the temperature gradient and reaction steps, which provides a beneficial reference for selecting
available sorbent. Taking the BeCl$_2$ as the example, Table 4.2 shows that the reaction
is divided into two steps: one is 2–4 coordination ions, and the other is 4–6 coordination ions. The temperature gradient reaches 310 °C between two steps, which is
very difficult for the application of this sorbent.

The equilibrium mechanism of solid chemisorption refrigeration can also be
expressed with the Clapeyron figure shown in Fig. 4.2 [53].

During the desorption stage, sorbents are heated by the heat source and sorb the
desorption heat (Q_1), and the refrigerant gas releases the condensation heat (Q_2) into
the condenser. In the sorption stage, sorbents are cooled by the heat sink and release
the sorption heat (Q_4). At the same time, the refrigerants evaporate into the reactor,
and the refrigerating capacity (Q_3) is achieved in the evaporator.

Table 4.1 The reaction parameters of halides and ammonia [52]

NO	Reaction	ΔH (J/mol)	ΔS (J/(mol K))	C_p (J/(mol K))
0	NH₃	23,366	150.52	80.27 (liquid)
1	Zn10-6	29,588	219.23	71.27
2	Cu10-6	31,387	227.72	71.81
3	Sn9-4	31,806	224.86	70.60
4	Pb8-3.25	34,317	223.76	70.05
5	Ba8-0	37,665	227.25	75.10
6	Sn4-2.5	38,920	229.82	70.60
7	Pb3.25-2	39,339	230.27	70.05
8	Ca8-4	41,013	230.30	72.52
9	Sr8-1	41,431	228.80	75.53
10	Ca4-2	42,268	229.92	72.52
11	Zn6-4	44,779	230.24	71.27
12	Pb2-1.5	46,035	230.89	70.05
13	Pb1.5-1	47,290	232.50	70.05
14	Mn6-2	47,416	228.07	72.86
15	Zn4-2	49,467	230.24	71.27
16	Cu5-3.3	50,241	230.75	71.81
17	Fe6-2	51,266	227.99	76.57
18	Cu3.3-2	56,497	237.22	71.81
19	Co6-2	53,986	228.10	78.41
20	Pb1-0	55,660	231.04	70.05
21	Mg6-2	55,660	230.63	71.31
22	Ni6-2	59,217	227.75	71.60
23	Ca2-1	63,193	237.34	72.52
24	Ca1-0	69,052	234.14	72.52
25	Mn2-1	71,019	232.35	72.86
26	Mg2-1	74,911	230.30	71.31
27	Fe2-1	76,167	231.91	76.57
28	Co2-1	78,134	232.17	78.41
29	Ni2-1	79,515	232.17	71.60
30	Zn2-1	80,352	229.72	71.27
31	Mn1-0	84,202	233.18	72.86
32	Fe1-0	86,880	233.01	76.57
33	Mg1-0	87,048	230.88	71.31
34	Co1-0	88,303	232.80	78.41
35	Ni1-0	89,810	233.01	71.60
36	Zn1-0	104,625	227.79	71.27

Table 4.2 The ammoniate halides with reasonable refrigeration performance [51]

Complex n-m	Energy Q_{cal}/W h	Step	Temperature gradient °C	Complex n-m	Energy Q_{cal}/W h	Step	Temperature gradient °C
MgCl$_2$/0-6	528	3	240	NiCl$_2$/2-6	272	1	0
NiCl$_2$/0-6	467	3	210	CaCl$_2$/2-8	270	2	10
CoCl$_2$/0-6	444	3	240	SrCl$_2$/1-8	266	1	0
FeCl$_2$/0-6	431	3	240	CoCl$_2$/2-6	258	1	0
CaCl$_2$/0-8	409	4	200	NH$_4$Cl/2-6	251	1	0
MnCl$_2$/0-6	408	3	270	AlCl$_3$/2-6	250	1	0
BeCl$_2$/0-6	324	2	310	LiCl/2-5	249	3	70
NaCl/0-5	317	1	0	FeCl$_2$/2-6	244	1	0
SrCl$_2$/0-8	308	2	30	CuCl$_2$/2-6	235	3	30
MgCl$_2$/2-6	306	1	0	LiCl/0-1	235	1	0
CdCl$_2$/0-6	305	3	230	BaCl$_2$/0-8	234	1	0
NH$_4$Cl/0-3	279	1	0	HgCl$_2$/0.67-9.5	231	3	280
CuCl/0-3	278	3	180	MnCl$_2$/2-6	227	1	0
BeCl$_2$/2-4	275	1	0				

Fig. 4.2 The Chapeyron figure for solid chemisorption: Point 1, desorption; Point 2, condensation; Point 3, evaporation; Point 4, sorption [53]

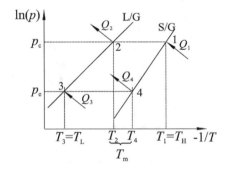

Even for the same working pair, the equilibrium Clapeyron line is not sole, which is related to different reaction processes [30]. With the Clapeyron figure, the pseudo-sorption equilibrium phenomenon [30, 32] has been developed, where the zero reaction rate area is shown between lines B and C in Fig. 4.3.

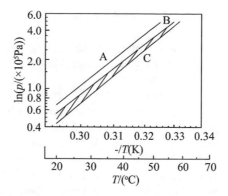

Fig. 4.3 Pseudo sorption equilibrium phenomenon [30]

4.1.2 Precursor State of Solid Chemisorption in Halide-Ammonia

Since the effective distance of Van der Waals forces for physical adsorption is much larger than that of chemical bonds for solid chemisorption, physical adsorption takes place before solid chemisorption for the chemical sorbents, which is named as a precursor state of solid chemisorption [54]. The precursor state of the solid chemisorption phenomenon is shown in Fig. 4.4 [54, 55].

Figure 4.4 shows that the precursor state of solid chemisorption has a noticeable impact on the refrigeration process. If the distance between the molecules of halide and ammonia is reduced (from d_2 to d_1), the activation energy of the solid chemisorption reaction will decrease from E_{a2} to E_{a1}, meaning that it will be easier to transit from physical adsorption to solid chemisorption.

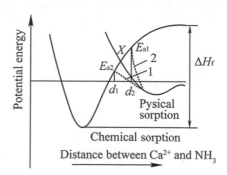

Fig. 4.4 Precursor of solid chemisorption for calcium chloride-ammonia [54, 55]

4.1.2.1 Chemical Sorbent with the Different Expansion Space

Chemical sorbents generally have a large cycle sorption capacity. One typical design of the sorption bed is shown in Fig. 4.5 for the research on halide sorbent for different expansion space.

The sorbent bed (Fig. 4.5) consists of two half-part columns that are fully welded between the sorption bed and fin, ensuring high heat transfer performance. Two half-part columns are connected with flange seals. Flange utilizes convex and groove to avoid the leakage. Sorption bed and other experimental devices are connected by the flange. A folding is added at the edge of the fin to prevent the sorbent from falling from the fin. Mass transfer is located in the channel between fin and fin. For $CaCl_2$, since the thickness of chemical sorbent on the fin is different, expansion space and the gap between Ca^{2+} and Ca^{2+} are different, which will influence the chemical sorption precursor state.

Corresponding to the design in Fig. 4.5, when the volume ratio of expansion space and sorbent is 5:1, the state of ammoniate $CaCl_2$ after sorption is shown in Fig. 4.6. Because there are still at least 2 mol ammonia coordinates in 1 mol sorbent after desorption, the sorbent is swelling and full between fins after desorption. The ammonia coordinated in sorbent increases in the sorption process, causing the sorbent swelling obviously and entering in the mass transfer channel after sorption. The distribution of sorbent can be assumed as full between fins in the process of sorption and desorption as the mass transfer space is tiny, which is only about 5.8% of effective space for sorbent.

The ratio of expansion space and the volume of the sorbent is defined as r_{as}, which are 5:1 (sample 1), 3:1 (sample 2), 2:1 (sample 3) and 1.4:1 (sample 4), respectively. For $CaCl_2$, it is assumed that the distribution of Ca^{2+} of sample 1 in the sorption process is shown in Fig. 4.7a. The distances between Ca^{2+} for sample 2, sample 3, and sample 4 are respectively 1.5, 2, and 2.5 times that of sample 1, according to different r_{as}, which is shown in Fig. 4.7b–d shows that the distribution of Ca^{2+} is different for the sorbents with different values of r_{as}. The distribution of Ca^{2+} is loose for sample 1, under which condition the shield factor of NH_3 is large because the concentration of NH_3 is large. Thus, it will be difficult for NH_3 to enter the effective

Fig. 4.5 Bulk sorbent bed
[56]

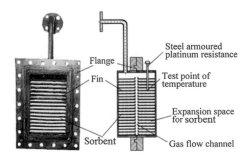

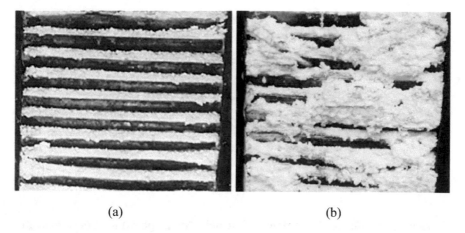

(a) (b)

Fig. 4.6 States of ammoniate CaCl$_2$ after sorption and after desorption [57]. **a** After desorption; **b** After sorption

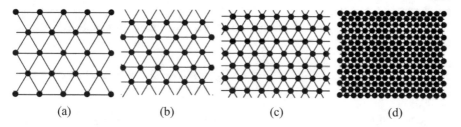

(a) (b) (c) (d)

Fig. 4.7 Distribution of Ca^{2+} for sorbent in the process of sorption. **a** Sample 1 (r_{as} is 5:1); **b** Sample 2 (r_{as} is 3:1); **c** Sample 3 (r_{as} is 2:1); **d** Sample 4 (r_{as} is 1.4:1) [57]

reaction range of Ca^{2+}. The distance between Ca^{2+} is too short for sample 4. Thus, the permeation of NH$_3$ will be influenced by low evaporation pressure.

4.1.2.2 Attenuation Performance of the Sorbent and Its Chemical Sorption Precursor State

Since the distribution of the molecules will be different when the expansion space of sorbent is different, the molecular sorption amounts of sorbents under the condition of different r_{as} is as follows:

$$N_g = \frac{x \times m_C}{m_N} \tag{4.5}$$

where m_c is a molecular mass of halide, m_N is the molecular mass of NH$_3$.

(a)   (b)   (c)

Fig. 4.8 Sorbents after sorption and desorption [21]. **a** r_{as} is 5:1; **b** r_{as} is 2:1; **c** r_{as} is1.4:1 [57]

When r_{as} is different, the agglomeration and swelling phenomena of sorbent after desorption and sorption will be different. When r_{as} is large, the swelling phenomenon will be severe. When r_{as} is small, the severe expansion will lead to agglomeration of the sorbent. Pictures of $CaCl_2$ with different r_{as} after the sorption and desorption are shown in Fig. 4.8.

The reaction formulas for the multiplex reaction between $CaCl_2$–NH_3 are as follows:

$$CaCl_2 \times 8NH_3 + \Delta H_1 \leftrightarrow CaCl_2 \times 4NH_3 + 4NH_3 \quad \text{at the temperature of } T_{e1} \tag{4.6}$$

$$CaCl_2 \times 4NH_3 + \Delta H_2 \leftrightarrow CaCl_2 \times 2NH_3 + 2NH_3 \text{ at the temperature of } T_{e2} \tag{4.7}$$

$$CaCl_2 \times 2NH_3 + \Delta H_3 \leftrightarrow CaCl_2 + 2NH_3 \text{ at the temperature of } T_{e3} \tag{4.8}$$

where ΔH_1, ΔH_2, and ΔH_3 are enthalpies of transformation for reactions (J/mol), and T_{e1}, T_{e2}, and T_{e3} are equivalent temperatures for reactions.

For 2 mol ammonia complex, the complex formed is the linear mode through sp orbital. For $CaCl_2 \cdot 4NH_3$, the complex formed is the tetrahedron mode through sp^3 hybrid orbital. Compared with $CaCl_2 \cdot 4NH_3$, for $CaCl_2 \cdot 6NH_3$ and $CaCl_2 \cdot 8NH_3$, the ammonia is occupied in d orbital, which is formed as regular octahedron and dodecahedron structure. When the sorbents are in the complexion process, the complex structure continuously adjusts from linear mode to dodecahedron mode [58]. For the ample expansion space, adjustment of sorbent pore influences the concentration change of ammonia around Ca^{2+}, thus will strengthen or weaken the repulsive force of anion. However, when the expansion space of the sorbent is limited, the sorbent tends to connect in the desorption and sorption process. This adjustment has little influence on the structure of sorbent, and sorption performance will be stable for further sorption. This phenomenon can be demonstrated from the performance attenuation curves of calcium chloride-ammonia working pair.

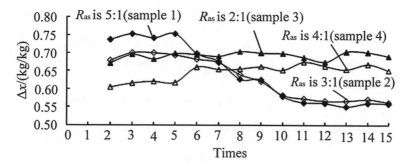

Fig. 4.9 Sorption performance attenuation curves [57]

Figure 4.9 shows that the performance of sample 1 is similar to that of sample 2. They both have performance attenuation and similar sorption capacity after the attenuation. The attenuation does not exist for sample 3 and sample 4, and sorption capacities become stable after the second sorption. The stable cycle sorption capacity of sample 3 is about 0.71 kg/kg. Contrarily, the cycle sorption capacity of sample 1 degenerates from 0.75 to 0.57 kg/kg; the most massive attenuation value is 31.6% comparing with the sorption capacity before attenuation. For sample 4, small expansion space leads to the severe agglomeration and limitation of forming $CaCl_2 \cdot 8NH_3$. Therefore, its sorption performance is lower than that of sample 3.

In the sorption process, sorption rate is faster than the desorption rate, and net sorption rate K_v is

$$K_V = \frac{d\theta}{dt} = K_a(1 - \theta)p - K_d\theta \qquad (4.9)$$

where θ is surface coverage, which is the ratio of sorption capacity x and the largest sorption capacity x_{max}. According to the Arrhenius activated energy [59, 60], the constant of reaction rate K_a in sorption and the reaction rate K_d in desorption are:

$$K_a = A_{fe} \exp \frac{-E_a}{RT} \qquad (4.10)$$

$$K_d = A_{fe} \exp \frac{-E_d}{RT} \qquad (4.11)$$

where A_{fe} is an anterior factor, and R is the universal gas constant. E_a is sorption activated energy, E_d is desorption activated energy [51]. Use Eqs. 4.10 and 4.11 to substitute K_a and K_d in Eq. 4.9, and then make a logarithmic transformation:

$$\ln K_V = \ln A_{fe} - \frac{E_a}{RT} + \ln\left[(1 - \theta)p - \theta \exp(\frac{\Delta H_r}{RT})\right] \qquad (4.12)$$

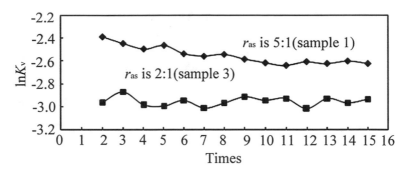

Fig. 4.10 The ln K_v for different experiments [57]

where anterior factor A_{fe} is seemed like a constant, and ΔH_r is the sorption heat. E_a is inversely proportional to the net sorption rate. Two typical curves, average ln K_v for the attenuation curves of sample 1 and sample 3 are shown in Fig. 4.10.

ln K_v of sample 3 does not change very much in repeated experiments; i.e., activate energy does not change very much according to the relation between E_a and ln K_v. It is resulted in the distance between Ca^{2+} and NH_3. The distance between Ca^{2+} and NH_3 is limited and does not change very much in sorption, thus the required activated energy will be stable according to the chemical sorption principle in Fig. 4.4. ln K_v of sample 1 in Fig. 4.10 decreases in the experiments of anterior ten times, and it becomes stable from 11th experiment; i.e., the activated energy increases for repeated ten times of experiments and becomes stable after the 11th experiment. This result is also coincident with the chemical sorption principle (Fig. 4.4). The distance between Ca^{2+} and NH_3 continually increases in the experiments of anterior for ten times for sample 1 because r_{as} is vast, and there is enough expansion space, thus the activated energy that is required for the transition from precursor state to chemical sorption increases. In the 11th experiment, sample 1 is filled between fins because of swelling. Then the distance between NH_3 and Ca^{2+} does not change very much since the further swelling of the sorbent is limited by the space between fins, thus activated energy becomes stable after the 11th experiment.

4.1.2.3   Isobaric Sorption Performance and Activated Energy

Activated energy is constant if the temperature is a constant, and the anterior factor is also a constant [61, 62]. Differentiate ln K_v to T in Eq. 4.12, and the result of activated energy is shown in Eq. 4.13.

$$E_a = \left[\frac{1}{K_v} \frac{dK_v}{dT} - \frac{\Delta H_r}{RT^2} \exp\left(\frac{\Delta H_r}{RT} \right) \frac{1}{(1-\theta)p - \theta \exp\left(\frac{\Delta H_r}{RT} \right)} \right] T^2 R \qquad (4.13)$$

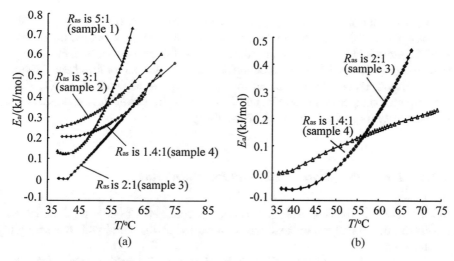

Fig. 4.11 Activated energy of different samples [57]. **a** Evaporation pressure is 430 kPa; **b** Evaporation pressure is 595 kPa

In the process of sorption refrigeration, activated energy required is negative, and the higher the absolute value is, the better the cooling effect is. To make the results of the activated energy more intuitive, the absolute value of activated energy is generally used. For the halide-ammonia working pairs, when evaporation pressure is 430 kPa, the activated energy of sorbent with different r_{as} is calculated and shown in Fig. 4.11a. In the sorption process, activated energy decreases when the temperature increases, which is mainly related to the stability and instability constant for the sorption/desorption process. The sorption reaction is exothermic reaction and stability constant will increase when the temperature decreases [63]. In Fig. 4.11a, average activated energy for sample 3 and sample 4 is lower than that of sample 1 and sample 2, which is in accordance with Fig. 4.4. Ca^{2+} distribution of sample 3 and sample 4 are concentrated. Thus the longest distance between Ca^{2+} and NH_3 gas molecule is within the normal range for chemisorption, for which activated energy should be less than that of sample 1 and sample 2 considering the shield factors. In Fig. 4.11a, activated energy for sample 4 is different from the chemisorption theory. For chemisorption theory, when r_{as} decreases, the distance between Ca^{2+} and NH_3 molecule is influenced by the concentrated distribution of Ca^{2+}, which should be within the range of Ca^{2+}. But in Fig. 4.11a, activated energy of sample 4 is higher than that of sample 3. This is due to the mass transfer problem in the process of sorption, which leads to lousy permeability inside the sorbent, reducing the sorption rate. Increasing evaporation pressure to 595 kPa, thereby pressure difference between the refrigerant and sorbent, is increased shown in Fig. 4.11b. At this time, activated energy of sample 4 and sample 3 are similar, which shows that chemisorption is influenced both by the chemisorption precursor state and mass transfer performance. If the distance between the complex molecules is not small enough to affect the mass transfer, the chemisorption precursor state determines the chemisorption. If the distance between the complex molecules is too small and the saturated pressure of the refrigerant is high enough to solve the problem of mass transfer, both chemisorption precursor state and mass transfer performance is essential for the chemisorption process. According to the principle of activated energy, as shown in Fig. 4.11, when the evaporation pressure is 430 kPa and 595 kPa, the best molecular distance is for the sample with r_{as} of 2:1 and 1.4:1, respectively.

4.1.3 The Desorption Hysteresis Phenomenon

In early research, the chemisorption reaction is considered to be reversible, which means the desorption parameters can be regarded as the same with the sorption parameters.

However, as discussed in the Chap. 3, halide-ammonia working pairs exhibit the desorption hysteresis phenomenon [64–66], which has been validated but cannot be explained by the capillary condensation theory. The main reason is the influence of activated energy on sorption/desorption, which has been analyzed in Chap. 3. Such a phenomenon will influence the kinetic characteristics and models of solid sorbents.

4.2 Analysis of Five Classical Kinetic Models

4.2.1 An Analogical Model Considering Various Classifications of Kinetic Parameters

Spinner and Lebrun tried to develop a quickly resolved model [39], which could be used to predict the behavior of chemical heat pumps (CHPs) and find out the best operation mode. An analogical model was chosen instead of the phenomenological model, which did not consider the properties and mechanism of the reaction medium. As the whole reaction includes two stages, the chosen model representing the sorption process is shown in Eqs. 4.14–4.17:

$$\frac{dX}{dt} = (1 - X)^{n_1} C_1 \exp(\frac{-E_1}{T}) \ln \frac{p}{p_{e_1}} \tag{4.14}$$

$$\frac{dY}{dt} = (1 - Y)^{m_2} X^{n_2} C_2 \exp(\frac{-E_2}{T}) \ln \frac{p}{p_{e_2}} \tag{4.15}$$

$$\frac{dN}{dt} = 2(\frac{dX}{dt} + \frac{dY}{dt}) \tag{4.16}$$

$$mC_p \frac{dT_{salt}}{dt} = n_{salt}\Delta H \frac{dN}{dt} + U_p A(T_c - T_{salt}) \tag{4.17}$$

where X is the conversion degree (x_1/x_{1max}) of reaction I $(Ca(CH_3NH_2)_2Cl_2 + 2CH_3NH_2 \Longleftrightarrow Ca(CH_3NH_2)_4Cl_2)$, Y is the conversion degree (x_2/x_{2max}) of reaction II $(Ca(CH_3NH_2)_4Cl_2 + 2CH_3NH_2 \Longleftrightarrow Ca(CH_3NH_2)_6Cl_2)$, N is the number of moles of gas sorbed per mole sorbent and U_p is the corresponding global heat exchange coefficient.

Spinner and Lebrun [39] refer that it is better to determine each parameter independently to improve the accuracy of parameters. Thus, the kinetic parameters are divided into three groups. The first group consists of the pseudo-energies of activation, E_1, and E_2, which are directly determined from those obtained by Mazet. The second group is the kinetic coefficients, consisting of C_1 and C_2. The third group represents the pseudo-orders of reaction, including n_1, n_2, and m_2, which need to be determined first and are considered as the external parameters.

The simulation of kinetic parameters shown in Eqs. 4.14 and 4.15 are inspiring, but some confusing aspects still exist. One of these lies in understanding how the pseudo-orders of reaction can be calculated when the kinetic coefficients are still unknown. Lebrun and Spinner [39] derived the value of n_1 as 1.5, but n_2 and m_2 were both equal to 1. However, the pseudo-orders should be obtained by appropriate experimental simulation (instead of a uniform relative value) since each experiment has its pseudo-orders. The second point is that since the reaction temperature difference between the two stages is very small, it is difficult to obtain the exact difference between the parameters of different stages. Thirdly, the pseudo-energies of activation of all the experiments cannot be assumed as the same. This is because the pseudo-energies of activation will change when the concentration of the reaction medium changes [67]. As a result, X and Y do not correspond with the experimental data. Therefore, the application of this method is the right direction but still needs to be improved.

4.2.2 An Analogical Model Uncoupling the Kinetic and Thermal Equation

It was proposed that it is possible to separate the kinetic and thermal parameters existing in the transformation model by some suitable procedures [30]. It is easy to understand that during the reaction, the kinetic reaction process coupled with the heat and mass transfer leads to significant gradients in both temperature and conversion, as well as a vast difference between the local thermodynamic conditions and the constraint value. What is more, the coupling process adds the difficulty in solving the equations. Therefore, Mazet et al. [30] developed a new approach to uncouple the two types of coefficients, which is possible if knowing temperature profiles from the experimental data during the reaction. Using this method, the kinetic equation can be independently solved without considering the thermal equations. Once the local kinetics is available, it is easier to calculate the local temperatures and advancements as well as to obtain the thermal coefficients. The apparent advantage of this two-stage method lies in reducing the number of parameters to be ensured. Furthermore, it helps to research these two phenomena separately. No matter which kinetic law is chosen, it would not affect the heat transfer phenomena and the thermal equations.

This method gives us an essential reference for solving this kind of thermodynamic question. The reaction and kinetic equations are shown in Eqs. 4.18–4.24. It needs to be mentioned that x represents the process R_1, while y represents the process R_2.

$$R_1: CaCl_2(6CH_3NH_2) \Leftrightarrow CaCl_2(4CH_3NH_2) + 2CH_3NH_2 \qquad (4.18)$$

$$R_2: CaCl_2(4CH_3NH_2) \Leftrightarrow CaCl_2(2CH_3NH_2) + 2CH_3NH_2 \qquad (4.19)$$

$$\frac{dx}{dt} = f(x) \cdot k(p, T) \qquad (4.20)$$

$$f(x) = (1 - x)^{m_x} \qquad (4.21)$$

$$f(y) = [(1 - y)x]^{m_y} \qquad (4.22)$$

$$k(p, T) = s \cdot \exp(-\frac{E}{RT}) \cdot f'(p, T) \qquad (4.23)$$

$$f'(p, T) = (p_c - p_{eq}(T))/p_c \qquad (4.24)$$

However, some deficiencies still exist in Eqs. 4.18–4.24. Firstly, Mazet et al. report that at the zone level (global level), the second reaction can start ($y > 0$) even if the first one is not yet completed ($x < 1$). However, this leads to some confusion since the model is described by

$$\frac{dx(t, r)}{dt} = \{[1 - x(t, r)]\}^{m_x} Ar_x \frac{p_c - p_{eqx}[T(t, r)]}{p_c} \qquad (4.25)$$

$$\frac{dy(t, r)}{dt} = \{[1 - y(t, r)]x(t, r)\}^{m_y} Ar_y \frac{p_c - p_{eqy}[T(t, r)]}{p_c} \qquad (4.26)$$

which looks more like the local model instead of the global model. However, in the local model, the second reaction can start only when the first reaction is completed. Therefore, the definition of m_Y needs more discussion. The second confusing aspect is due to considering the Arrhenius term as a constant; thus, the variation can be ignored compared with the non-equilibrium term. However, for some reaction mediums

Table 4.3 Simulation results for parameters under various constraint pressure and temperature [30]

p_c	T_c °C	Ar_x 10^3/s	m_x	Ar_y 10^3/s	m_y	ΔX_{gl}	$\Delta \dot{X}_{gl}$ $\times 10^4$
2.8	69.7	0.415	2.13	0.846	1.10	0.0228	0.208
1.85	63.8	0.300	1.7	0.460	0.65	0.0177	0.196
	56.5	0.680	1.63	0.156	0.63	0.0078	0.252
	49.4	0.687	2.66	0.103	0.20	0.0088	0.134
1.15	55.1	0.365	1.5	0.226	0.70	0.0139	0.181
	48.3	0.400	1.9	0.133	0.40	0.0092	0.077
	41.4	0.374	1.61	0.081	0.90	0.0120	0.146
0.7	44.2	0.350	1.17	0.146	1.0	0.0128	0.178
	37.2	0.585	1.71	0.085	1.0	0.0074	0.152

whose operation temperature range is broad enough, the Arrhenius term changes significantly during the reaction, showing that it cannot be assumed to be a constant.

When it comes to the simulation of parameters, the result is not good enough due to sharp fluctuations in their values. The details are shown in Table 4.3 [30], where $\Delta X_{gl\,abs}$ is the global sorption capacity and $\Delta \dot{X}_{gl}$ is the global sorption rate. The reason for such fluctuations is explained because of the notable variation in T_c for a given p_c. In other words, as long as the constraint temperature changes, the parameters may change. This indicates two main points: first, it is almost impossible to obtain stable parameters under various conditions; second, it is worth finding out the relationship between the constraint conditions and the variation in the parameters, even though Mazet et al. did not implement it.

4.2.3 An Analytical Model Based on the Thermophysical Properties of the Reaction Medium

Generally, when designing a reactor, it is necessary to conduct dynamic simulations, using either local or global models. Initially, the characteristic of the heat and mass transfer, the geometry of the reactor, and the operating conditions can be chosen. Then, the response at the global level and the thermal power of the reaction under the given constraint, ΔT_{eq}, can be obtained. Since simulation needs successive iterations of parameters to reach the ideal power, the numerical calculations are very complex, which is not suitable for the optimization and pre-dimensioning of the system.

A new analytical model has been put forward to simplify the dimensioning procedure and promote its reliability [29, 38]. It links with the performances and necessary parameters and establishes a relationship among geometry of the reactor, the desired average thermal power, the specific working conditions, and thermal characteristics of the reaction medium. Therefore, the necessary parameters can be determined by

Fig. 4.12 Principle diagram
for determining the working
conditions [29]

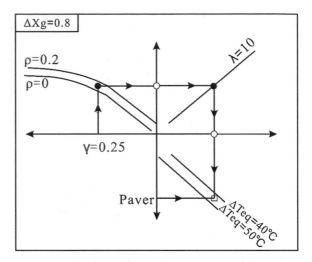

considering the required performance of the system, as shown in Fig. 4.12 [29].
For instance, combining the objective thermal power with the constraint parameters
(thermal characteristic value and geometry value) can be used to deduce the unknown
parameters. Similarly, we can optimize the parameters one by one by fixing other
parameters until the best global properties are obtained.

The simplified specific models are given in Eqs. 4.27 and 4.28.

$$\frac{\lambda}{r}\frac{d}{dr}(r\frac{d\langle T(r)\rangle}{dr}) - \nu n_{salt}\Delta H_r\frac{\Delta X(r, t_r)}{t_r} = 0 \tag{4.27}$$

$$\Delta X(x, t) = \frac{\theta(r, t)}{\theta_i} \text{ with } \theta(r, t) = \frac{T(r, t) - T_{eq}}{T_c - T_{eq}} \tag{4.28}$$

where λ is the effective thermal conductivity of the fixed-bed reaction medium, ν is
the coefficient of the chemical reaction, T_c is the constraint temperature applied to
the reactor, $\Delta X(r, t_r)$ is the local extent of the reaction, t_r is the reaction time corresponding to the given global extent, $\theta(r, t)$ is the local dimensionless temperature,
θ_i is the local dimensionless temperature when the reaction is locally finished.

It is evident that the simplified analytical models just consider the heat transfer
process, regardless of the mass transfer process and the inner kinetic reaction. Moreover, it needs to be mentioned that they linearize the relationship between the local
temperature and local extent directly through the patterns obtained from the experimental data. Using this method, the width of the reaction front and the profile of
local extents can be simulated easily.

Even though this model can decrease the application threshold due to its reliable
simplification, it has some disadvantages. First, it does not consider the mechanism
of the reaction and thus may become invalid if the reaction medium changes, since the

relationship between temperature and the local extent may be difficult to determine. Second, when the limiting factor is no more heat transfer but the mass transfer or chemical reaction, the model cannot be applied. However, when the reaction mechanism is not precise, this model can be used to design the dimensions of the reactor in chemisorption systems.

4.2.4 Phenomenological Grain-Pellet Model

Goetz proposed a phenomenological model [32], which is still one of the most classical kinetic models. The concept of grain-pellet is used because it makes possible to define two different dimensions of the reaction medium. The grain is the basic unit where the reaction takes place, while the pellet is the combination of reactive grains. The model considers all types of restrictions, including heat and mass transfer through the reaction medium, and chemical reactions occurring at the reaction interface. The details of the model are listed in Table 4.4 [32].

In Table 4.4, N_g is the molar sorption capacity, r_g and r_c are the radii of the crystalline grain and reaction surface respectively, T_c is the constraint temperature, K_s is the gas permeability, p_i is the pressure on the reaction surface, p_c is the constraint pressure, p_e is the equilibrium pressure, V_m is the molar volume, K and M are kinetic parameters and x is the sorption rate. Sorption is represented by 'a' while desorption is represented by 'd'.

Lu et al. simplified the continuous reaction interface into a sharp interface and established a model that considers heat and mass transfer instead of the chemical reaction process [40]. Later, Wang added the term of cation distance, which influences the kinetic process and mass transfer on a crystalline level [69]. This phenomenological model is unique due to its comprehensiveness. But few researchers choose it for simulating their systems because of its complexity.

Table 4.4 Equations for the consumption of a reactive grain [32]

	Sorption	Desorption
Mass transfer	$\frac{dN_g}{dt} = \pm \frac{r_g r_c}{r_g - r_c} \frac{4\pi}{RT_c} K_s(p_c - p_i)$	
Chemical kinetics	$\frac{dN_g}{dt} = 4\pi r_c^2 K_a \left(\frac{p_i - p_{ea}(T_c)}{p_{ea}(T_c)} \right)^{M_a}$	$\frac{dN_g}{dt} = 4\pi r_c^2 K_d \left(\frac{p_{ed}(T_c) - p_i}{p_{ed}(T_c)} \right)^{M_d}$
Variation in grain size	$r_g^3 = r_c^3 +$ $\left(r_{g[[Mn(NH_3)_2Cl_2]]}^3 - r_c^3 \right) \frac{V_{m[Mn(NH_3)_6Cl_2]}}{V_{m[Mn(NH_3)_6Cl_2]}}$	$r_g^3 = r_c^3 +$ $\left(r_{g[Mn(NH_3)_2Cl_2]}^3 - r_c^3 \right) \frac{V_{m[Mn(NH_3)_6Cl_2]}}{V_{m[Mn(NH_3)_2Cl_2]}}$
Advancement	$x = 1 - \left\{ \frac{r_c}{r_{g[Mn(NH_3)_2Cl_2]}} \right\}^3$	$x = 1 - \left\{ \frac{r_c}{r_{g[Mn(NH_3)_6Cl_2]}} \right\}^3$

4.2.5   A Numerical Model for CHPs

In order to propose an appropriate kinetic model for the behavior of the reactor, Castaing and Neveu [36] compared simulation results based on three classical kinetic laws established by Lebrun [39], Mazet [30], and Goetz [32] with the experimental results obtained by Setaram through DSC. In the Clapeyron diagram, each point corresponds to four different reaction rates since the degrees of advancement (of both $NiCl_2$-6/$2NH_3$ and $MnCl_2$-6/$2NH_3$) have four values ($X = 0.2, 0.4, 0.6, 0.8$) for the experimental data and each model. Thus, if the kinetic behavior needs to be represented generally, several similar figures are required at each constraint temperature. In order to make the comparison for kinetic behaviors simpler and clearer, a new parameter, the Carnot temperature, was defined to indicate the driving force and the non-equilibrium state, as shown in Eq. 4.29

$$\theta = 1 - \frac{T_c}{T_{eq}(p_c)} \tag{4.29}$$

where θ is related to the T_c and p_c constraint conditions, which makes it possible to show all the experimental analysis in a single figure.

Through this approach, all the experimental points can be shown in a single figure, and the comparison results indicate that Goetz's model is much better than the other two kinetic laws (considering the relative error from the experimental data). However, the calculation time increases by 2–5 times. After comparing other kinetic laws, a novel idea to set up a new kinetic model for the solid–gas thermochemical reaction was proposed [36]. In this case, the Carnot temperature is replaced by a new corrected value, and a global model is obtained

$$\theta^* = \theta \cdot 1 - X \tag{4.30}$$

$$\frac{dX}{dt} = \left(\frac{U_{sw} \cdot S_{sw} \cdot T_{eq}(p_c)}{n_s \cdot v \cdot \Delta H_r^0} \right) \cdot \theta \cdot (1 - X) \tag{4.31}$$

where θ^* is the corrected Carnot temperature, U is the global exchange coefficient, S is the exchange surface, n_s is the number of moles of salt, and v is the volume. The relationship between the corrected Carnot factor and the reaction rate is shown in Fig. 4.13 [36].

This model makes the reactor linear and simplifies the simulation. However, some limitations still exist. First, it constructs a global model by considering that the heat exchange is equal to the latent heat of the thermochemical reaction, which is not accurate enough. Second, the variation range of T_{eq} cannot be extensive, which limits the model's application. Third, the linear relationship may not be suitable in different kinds of reactors due to the influence of different reaction media and

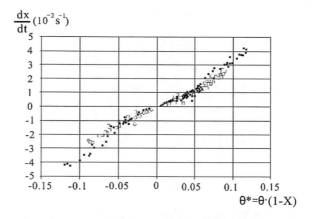

Fig. 4.13 Reaction rates in terms of the θ^* constraint parameter for all the experimentation carried out on the MnCl$_2$/NH$_3$ pair at $X = 0.2$ (■), $X = 0.4$ (□), $X = 0.6$ (◆), $X = 0.8$ (◇) [36]

heat and mass transfer. Meanwhile, it emphasizes the regular pattern of the data and phenomena alone, instead of the reaction mechanism itself.

4.3 Sorption Phenomena in Halide-Ammonia Working Pairs Developments in Recent years

4.3.1 Non-equilibrium Clapeyron Figure

According to Gibbs phase rule, for a system with k components, f phases, and r independent reactions, the number of degrees of freedom v can be calculated as $v = k + 2 - f - r$ [68]. Under this criterion, the halide-ammonia system should be mono-variant. Therefore, in early research, the chemisorption was considered to be a mono-variant controlling reaction under equilibrium conditions, which means that the reaction starts and finishes at the same state (pressure, temperature) in the Clapeyron figure. The theoretical performance of a chemisorption refrigeration system is often simulated based on this mechanism. However, Trudel et al. [69] studied a CoCl$_2$-NH$_3$ working pair and found out that the equilibrium Clapeyron curves are more complex than a single equilibrium line. Thus, the theory of pseudo-balance was proposed. In the pseudo-balance region, the system is bi-variant rather than mono-variant. Later, Aidoun and Ternan [70] further studied the instability and pseudo-steady state of transformation. Moreover, most of the practical processes occur under non-equilibrium conditions. The non-equilibrium Clapeyron figure was studied by Zhou et al. [71] using MnCl$_2$–NH$_3$, CaCl$_2$–NH$_3$, and NH$_4$Cl–NH$_3$ working pairs under non-equilibrium conditions, as shown in Fig. 4.14 [71].

The results show that non-equilibrium chemisorption is bi-variant, i.e., both pressure and temperature need to be specified to determine the reaction point. It is found that the more ammonia the halide sorbs, the further the experimental line deviates from the theoretical one. The explanation for the bi-variant phenomenon has

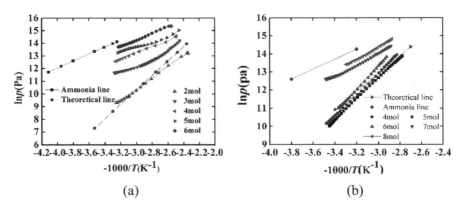

Fig. 4.14 The experimental and theoretical lines: **a** MnCl$_2$–NH$_3$, **b** CaCl$_2$–NH$_3$ [71]

been proposed by Zhong et al. [66]. The micro-crystallites of the sorbent are not homogenous due to the range of crystallite sizes inside the pores. Each specific micro-crystallite undergoes mono-variant decomposition or synthesis at the specific pressure and temperature. In contrast, the whole system behaves as bi-variant due to the non-uniform micro-crystallites [66, 72].

4.3.2 New Discoveries Related to Sorption Hysteresis

Since the sorption hysteresis phenomenon has been discovered, studying the desorption process has become difficult as it cannot be directly researched by analyzing the sorption process. Zhou et al. [71] studied three kinds of working pairs, NH$_4$Cl–NH$_3$, CaCl$_2$–NH$_3$, and MnCl$_2$–NH$_3$, and part of the results can be seen in Fig. 3.26. It is found that the hysteresis is more obvious when the sorbent needs higher desorption temperature. It is mainly related to the desorption heat, which depends on the activated energy for the desorption process. The desorption activation energy is the sum of the reaction heat and the sorption activation energy, and in which the reaction heat is much larger than the sorption activation energy. Thus for the MnCl$_2$–NH$_3$ working pair, the desorption activation energy is highest, followed by CaCl$_2$–NH$_3$ and then NH$_4$Cl–NH$_3$ working pair. Such a result coordinates with the hysteresis conditions of different working pairs.

Further analyzing the mechanism of sorption hysteresis for MnCl$_2$–NH$_3$ in Fig. 3.26, it is found that the severe desorption hysteresis occurred with 4 mol NH$_3$/mol MnCl$_2$. Such a phenomenon can be explained by the crystal field theory of ammoniate MnCl$_2$. The energy level of five orbits is different for ammoniate MnCl$_2$ by the action of the surrounding ligands degeneration. Such a level splitting phenomenon brings the extra stabilization energy, i.e., crystal field stabilization energy, under the condition that d orbits are not full. For MnCl$_2$–2NH$_3$, the ligands occupy sp orbit, for $MnCl_2$–$4NH_3$ the ligands occupy sp_3 orbit, and for $MnCl_2$–$6NH_3$ the ligands start occupying d orbit [73]. Thus the desorption process from $MnCl_2$–$6NH_3$ to $MnCl_2$–$4NH_3$ needs more energy, resulting in serious hysteresis. Likewise, the desorption hysteresis is serious for the desorption process from $CaCl_2$–$8NH_3$ to $CaCl_2$–$4NH_3$.

For the sorption/desorption process with double variables, the characteristics of the working pair can be analyzed by the D-A equation amended by Critoph [74]. Because the experimental Clapeyron curves of halides are bivariant, the D-A equations are applicable. The experimental isobaric sorption process is close to the experimental Clapeyron curve. For different working pairs, the D-A equation is:

$$x = x_0 \exp\left[-k\left(\frac{T}{T_{ref}} - 1\right)^n\right] \tag{4.32}$$

where x is sorption quantity (kg/kg), T_{ref} is refrigeration temperature, k, and n are constants for the sorbate. The constants in Eq. 4.32 are summarized in Table 4.5.

For the desorption process, the hysteresis makes the deviation of the isobaric desorption curves far from the experimental Clapeyron curves. The relationship between the hysteresis and the reaction heat is analyzed by the Arrhenius equation [75], as shown as follows.

$$\ln K = \ln A - \frac{E}{RT} \tag{4.33}$$

where K is the constant, A is the frequency factor, E is the activation energy of the reactants (kJ/mol). The complexion process between the chlorides and ammonia can be seen as Eq. 4.1 shown. Sorption velocity can be calculated by:

$$r_s = K_s \, M_a X_b \, (NH_3)_n^{c_1} (NH_3)^{c_2} \tag{4.34}$$

and K_s in the equation is:

$$\ln K_s = \ln A_s - \frac{E_s}{RT} \tag{4.35}$$

Desorption velocity is calculated by:

Working pairs	x_0	k	n
$MnCl_2$–NH_3	0.6911	3.345	1.713
$CaCl_2$–NH_3	1.2224	6.388	1.857
NH_4C–NH_3	0.668	40.9	1.578

Table 4.5 The coefficients of D-A equation for sorption working pairs

$$r_d = K_d M_a X_b (NH_3)_m^{c_3} \tag{4.36}$$

where K_d is:

$$\ln K_d = \ln A_d - \frac{E_d}{RT} \tag{4.37}$$

In the Eqs. 4.34–4.37, K_s, and K_d are constants, E_s is sorption activation energy (kJ/mol), E_d is desorption activation energy (kJ/mol). For equilibrium desorption condition, namely $r_s = r_d$:

$$K_s M_a X_b (NH_3)_n^{c_1} (NH_3)^{c_2} = K_d M_a X_b (NH_3)_m^{c_3} \tag{4.38}$$

$$K_d/K_s = M_a X_b (NH_3)_n^{c_1} (NH_3)^{c_2} / M_a X_b (NH_3)_m^{c_3} = K \tag{4.39}$$

The equilibrium constant is:

$$\ln K = \ln \frac{K_d}{K_s} = \ln A_1 - \frac{E_d}{RT} - \ln A_2 + \frac{E_s}{RT} = \ln A_0 - \frac{\Delta H}{RT} \tag{4.40}$$

$$\ln K = \ln x_{\text{(for desorption)}} - \ln x_{\text{(for sorption)}} \tag{4.41}$$

where A_0 is a constant, related to the type of metal chloride, x is sorption quantity (kg/kg). So the relationship between desorption hysteresis and reaction heat is as follows:

$$\ln K = \ln x_{\text{(for desorption)}} - \ln \left\{ x_0 \exp\left[-k\left(\frac{T}{T_{ref}} - 1 \right)^n \right] \right\} = \ln A_0 - \frac{\Delta H}{RT} \tag{4.42}$$

$$\ln x_{\text{(for desorption)}} = \ln \left\{ x_0 \exp\left[-k\left(\frac{T}{T_{ref}} - 1 \right)^n \right] \right\} + \ln A_0 - \frac{\Delta H}{RT} \tag{4.43}$$

Because the reaction heat of $MnCl_2$–NH_3 and $CaCl_2$–NH_3 are large, the sorption hysteresis phenomenon is serious. The reaction heat of NH_4Cl–NH_3 working pair is small enough, so the hysteresis can be neglected (both sorption and desorption curves are close to the experimental Clapeyron curves). Thus only the desorption hysteresis of $MnCl_2$–NH_3 and $CaCl_2$–NH_3 are analyzed. By Eq. 4.43, for manganese chloride A_0 is 18.02, and for calcium chloride, A_0 is 17.37. The comparison of the fitting results by Eq. 4.43 and the experimental results are shown in Fig. 4.15. Results show that when the reaction heat is more abundant, the simulation error will be smaller. For $MnCl_2$–NH_3, its reaction heat is 47.416 kJ/mol, and the most significant error between the experiments and the simulation is 3.1%. For $CaCl_2$–NH_3, the reaction heat is 41.013 kJ/mol, and the most significant error is about 9.8%.

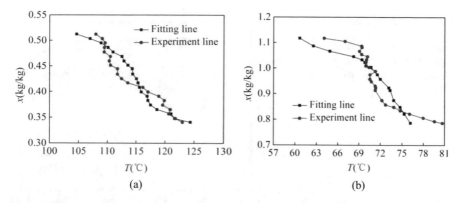

Fig. 4.15 Comparison between the experiments and Arrhenius equation: **a** MnCl$_2$–NH$_3$ working pair, **b** CaCl$_2$–NH$_3$ working pair [71]

Using this hysteresis model, the performance of a refrigeration cycle is analyzed. Under the condition of the heating process from 70 to 180 °C, 25 °C cooling temperature, and 0 °C refrigeration temperature, the refrigeration cycle with MnCl$_2$ is analyzed by theoretical equilibrium lines, fitted equations, and experimental results. COP is calculated as follows:

$$COP = \frac{Q_{ref}}{Q_h} = \frac{m \times \Delta x \times L_a - m \times \Delta x \times c_{am}(T_{en} - T_e)}{m \times \Delta x \times \Delta H + \int_{T_{en}}^{T_h} (mc_{ps} + m \times x \times c_{am})dT} \tag{4.44}$$

where Q_h is heating capacity (kJ), Q_{ref} is refrigerating capacity (kJ), Δx is the cycle sorption quantity of sorbent (kg/kg), L_a is the latent heat of ammonia (kJ/kg). T_e is the evaporation temperature (K), T_{en} is the environmental temperature (K), T_h is the highest heating temperature (K), c_{am} is the specific heat capacity of ammonia (kJ/(kg K)), c_{ps} is the specific heat of sorbent (kJ/(kg K)).

The analysis shows that the theoretical COP will become zero if the heating temperature is below 130 °C. If the heating source temperature is higher than 130 °C, the theoretical COP will be about 0.37 (Fig. 4.16). However, the experimental line shows that COP slowly changes from 0.287 to 0.352 as temperature increases. The error caused by the testing processes is mostly within 5%. The COP is also calculated by the D-A equation, and the results (Fig. 4.16) show that the fitting line agrees with the experimental line well, and most data fall within the ranges inside the error bars of experimental data. The maximum deviation is around 6.8%. However, if the results are analyzed by theoretical equilibrium lines, the maximum error can reach above 32%. The results in Fig. 4.16 indicate that the theoretical equilibrium line, which is extensively used for analyzing the refrigerating performance, is not applicable for the real non-equilibrium conditions since its error is more significant than 30%. The non-equilibrium lines and the relationship between desorption enthalpy and

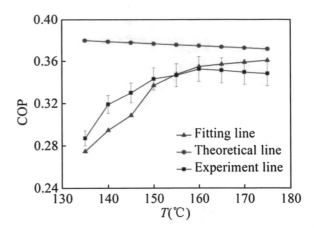

Fig. 4.16 Comparison among the experimental results, fitting line by D-A equation, and Arrhenius equation for MnCl$_2$–NH$_3$ [71]

desorption hysteresis should be considered for simulating the sorption refrigeration performance of halide-ammonia working pairs.

Gao et al. [76] attempted to solve the solution to the hysteresis by using novel compact composite multi-salt sorbents (NH$_4$Cl/CaCl$_2$/MnCl$_2$–NH$_3$), in which the sorption hysteresis is much less significant than that in the single halide, as shown in Fig. 3.30 [76]. An et al. proved that the hysteresis under non-equilibrium condition is related with the influence of both chemical reaction and heat transfer, and the sorption model and heat transfer model should be coupled to research its characteristics [77].

4.3.3 Part of the Composite Sorbent Models

Composite sorbents consisting of "salt inside porous matrix" [78, 79] are currently being preferred instead of pure chemical sorbents. For example, several working pairs have been used for icemakers, such as the composite LiCl/silica-methanol [79], LiBr/silica-water [80], CaCl$_2$/expanded graphite-ammonia [14], and BaCl$_2$/expanded vermiculite [66]. However, due to the additive matrix, the establishment of a theoretical kinetic model becomes much more complex. For several types of pure silica gel-water working pairs (FAM Z02, SAPO-34) [81, 82], the sorption/desorption processes obey the exponential equation, where the characteristic time is the function of the boundary temperatures. It is also found that (BaCl$_2$ + BaBr$_2$)/vermiculite and CaCl$_2$/silica-water working pairs obey the exponential equation [16, 83, 84], which is simpler than the D-A equation that is widely applied to explain physical adsorption. However, the mechanism of the exponential form is not clear. A single component sorbent-sorbate system has been derived from the viewpoint of thermodynamic requirements of chemical equilibrium, Maxwell relation, classical thermodynamics, Gibbs law, and energy banlance equations [85, 86].

4.4 The Development Direction of Kinetic Models for Further Research

Researchers pay little attention to the reaction kinetics of halide-ammonia working pairs but focus on the performance of experimental systems. This is the reason why almost no new kinetic models have been developed in the recent ten years to explain the halide-ammonia chemisorption. Some simplified typical models, such as the D-A equation, are being widely applied. The underlying chemisorption mechanisms and typical kinetic models proposed by different researchers have been analyzed to help other researchers obtain a clear understanding of the mechanism and modeling of halide-ammonia working pairs.

The difficulty of establishing a satisfactory model relies on two main aspects. The first one is that the chemisorption mechanism is complicated and is not very well understood. Therefore, most models cannot accurately depict the phenomenological model and are not able to reflect the reaction mechanism. The second one is that due to the influence of heat and mass transfer, the available models cannot be applied with wide ranges of reaction temperatures or pressures.

In our opinion, several other aspects need to be studied in the future.

(1) The isobaric sorption and desorption processes without the influence of heat and mass transfer. By testing such processes, the relationship between sorption capacity and temperature without the time term can be derived, which means that the kinetic model can be written in Eq. 4.45.

$$\frac{dx}{dt} = \frac{dx}{dT} \cdot \frac{dT}{dt} \tag{4.45}$$

This form can be much more universally applied because the term dx/dT is the same in different cases if the reaction medium is not changed. The only issue lies in researching the heat transfer conditions for obtaining dx/dt.

(2) The ligand field theory and activation energy theory should be paid attention to explain the chemisorption process accurately. They can contribute to confirm the relevant mechanism to explain sorption hysteresis and performance under non-equilibrium conditions.

References

1. Critoph RE, Metcalf SJ (2004) Specific cooling power intensification limits in ammonia-carbon adsorption refrigeration systems. Appl Therm Eng 24:661–678
2. Wang LW, Wu JY, Wang RZ, Xu YX, Wang SG (2003) Experimental study of a solidified activated carbon-methanol adsorption ice maker. Appl Therm Eng 23:1453–1462
3. Rezk ARM, Al-Dadah RK (2012) Physical and operating conditions effects on silica gel/water adsorption chiller performance. Appl Energy 89:142–149

4. Iloeje OC, Ndili AN, Enibe SO (1995) Computer simulation of a CaCl$_2$ solid-adsorption solar refrigerator. Energy 20:1141–1151
5. Goetz V, Spinner B, Lepinasse E (1997) A solid-gas thermochemical cooling system using BaCl$_2$ and NiCl$_2$. Energy 22:49–58
6. Libowitz GG, Feldman KT, Stein C (1997) Thermodynamic properties of metal hydrides for a novel heat pump configuration. J Alloy Compd 253:673–676
7. Ponec V, Knor Z, Černý S, Smith D, Adams NG (1974) Adsorption on solids
8. Lin S, Vannice MA (1991) Gold dispersed on TiO$_2$ and SiO$_2$: adsorption properties and catalytic behavior in hydrogenation reactions. Catal Lett 10:47–61
9. Lebrun M, Neveu P (1992) Conception, simulation, dimensioning, and testing of an experimental chemical heat pump. Ashrae Trans 98:420–429
10. Mauran S, Prades P, L'Haridon F (1993) Heat and mass transfer in consolidated reacting beds for thermochemical systems. Heat Recovery Syst CHP 13:315–319
11. Vasiliev LL, Mishkinis DA, Antukh AA, Kulakov AG, Vasiliev LL (2004) Resorption heat pump. Appl Therm Eng 24:1893–1903
12. Wang K, Wu JY, Wang RZ, Wang LW (2006) Composite adsorbent of CaCl$_2$ and expanded graphite for adsorption ice maker on fishing boats. Int J Refrig 29:199–210
13. Wang LW, Metcalf SJ, Critoph RE, Thorpe R, Tamainot-Telto Z (2011) Thermal conductivity and permeability of consolidated expanded natural graphite treated with sulphuric acid. Carbon 49:4812–4819
14. Jiang L, Wang LW, Jin ZQ, Wang RZ, Dai YJ (2013) Effective thermal conductivity and permeability of compact compound ammoniated salts in the adsorption/desorption process. Int J Therm Sci 71:103–110
15. Saha BB, Chakraborty A, Koyama S, Yu IA (2009) A new generation cooling device employing CaCl$_2$-in-silica gel-water system. Int J Heat Mass Transf 52:516–524
16. Okunev BN, Aristov Yu (2014) Making adsorptive chillers faster by a proper choice of adsorption isobar shape: comparison of optimal and real adsorbents. Energy 76:400–405
17. Langmuir I (1917) The constitution and fundamental properties of solids and liquids. Park I. solids. J Am Chem Soc 38:102–105
18. Gibbs JW (1957) The collected works of J. Willard Gibbs 1. Thermodynamics 6–591
19. Polanyi M (1932) Theories of the adsorption of gases: a general survey and some additional remarks 28
20. Dubinin MM (1960) The potential theory of adsorption of gases and vapors for adsorbents with energetically nonuniform surfaces. Chem Rev 60:235–241
21. Dubinin MM (1975) Physical adsorption of gases and vapors in micropores. Progress Surf Membrane Ence 9:1–70
22. Brunauer S, Emmett PH, Teller E (1938) Adsorption of Gases in Multimolecular Layers. J Am Chem Soc 60:309–319
23. Thu K, Chakraborty A, Saha BB, Ng KC (2013) Thermo-physical properties of silica gel for adsorption desalination cycle. Appl Therm Eng 50:1596–1602
24. Dubinin MM, Radushkevich LV, Dubinin MM, Radushkevich LV (1946) The equation of the characteristic curve of activated charcoal. Zhurnal Nevropatologii I Psikhiatrii Imeni Sskorsakova 79:843–848
25. Dubinin MM, Zolotarev PP, Nikolaev KM, Polyakov NS, Petrova LI, Radushkevich LV (1973) Investigation of the dynamics of adsorption in a broad range of breakthrough concentrations. Russ Chem Bull 21:1432–1437
26. Sellaoui L, Saha BB, Wjihi S, Lamine AB (2016) Physicochemical parameters interpretation for adsorption equilibrium of ethanol on metal organic framework: application of the multilayer model with saturation. J Mol Liq 233:537–542
27. Veselovskaya JV, Critoph RE, Thorpe RN, Metcalf S, Tokarev MM, Aristov Yu (2010) Novel ammonia sorbents "porous matrix modified by active salt" for adsorptive heat transformation: 3. Testing of "BaCl$_2$ /vermiculite" composite in a lab-scale adsorption chiller. Appl Therm Eng 30:1188–1192

28. Aristov Yu, Dawoud B, Glaznev IS, Elyas A (2008) A new methodology of studying the dynamics of water sorption/desorption under real operating conditions of adsorption heat pumps: experiment. Int J Heat Mass Transf 51:4966–4972

29. Stitou D, Crozat G (1997) Dimensioning nomograms for the design of fixed-bed solid-gas thermochemical reactors with various geometrical configurations. Chem Eng Process 36:45–58

30. Mazet N, Amouroux M, Spinner B (1991) Analysis and experimental study of the transformation of a non-isothermal solid/gas reacting medium. Chem Eng Commun 99:155–174

31. Sun LM, Meunier F (1991) An improved finite difference method for fixed-bed multicomponent sorption. AIChE J 37:244–254

32. Goetz V, Marty A (1992) A model for reversible solid-gas reactions submitted to temperature and pressure constraints: simulation of the rate of reaction in solid-gas reactor used as chemical heat pump. Chem Eng Sci 47:4445–4454

33. Lebrun M, Spinner B (1990) Simulation for the development of solid—gas chemical heat pump pilot plants Part I. simulation and dimensioning. Chem Eng Process 28:55–66

34. Lebrun M (1990) Simulation for the development of solid—gas chemical heat pump pilot plants Part II. simulation and optimization of Discontinuous and pseudo-continuous operating cycles. Chem Eng Process 28:67–77

35. Bjurström H, Suda S (1989) The metal hydride heat pump: Dynamics of hydrogen transfer. Int J Hydrogen Energy 14:19–28

36. Neveu P, Castaing-Lasvignottes J (1997) Development of a numerical sizing tool for a solid-gas thermochemical transformer—I. Impact of the microscopic process on the dynamic behaviour of a solid-gas reactor. Appl Therm Eng 17:501–518

37. Castaing-Lasvignottes J, Neveu P (1997) Development of a numerical sizing tool applied to a solid-gas thermochemical transformer—II. Influence of external couplings on the dynamic behaviour of a solid-gas thermochemical transformer. Appl Therm Eng 17:519–536

38. Stitou D, Goetz V, Spinner B (1997) A new analytical model for solid-gas thermochemical reactors based on thermophysical properties of the reactive medium. Chem Eng Process 36:29–43

39. Lebrun M, Spinner B (1990) Models of heat and mass transfers in solid—gas reactors used as chemical heat pumps. Chem Eng Sci 45:1743–1753

40. Lu HB, Mazet N, Coudevylle O, Mauran S (1997) Comparison of a general model with a simplified approach for the transformation of solid-gas media used in chemical heat transformers. Chem Eng Sci 52:311–327

41. Fan Y, Luo L, Souyri B (2007) Review of solar sorption refrigeration technologies: Development and applications. Renew Sustain Energy Rev 11:1758–1775

42. Fernandes MS, Brites GJVN, Costa JJ, Gaspar AR, Costa VAF (2014) Review and future trends of solar adsorption refrigeration systems. Renew Sustain Energy Rev 39:102–123

43. Sarbu I, Sebarchievici C (2015) General review of solar-powered closed sorption refrigeration systems. Energy Convers Manage 105:403–422

44. Chen C (2008) Review of solar adsorption refrigeration system for air conditioning. J Refrig 2008(4):1–7

45. Mas D, Randip K (2015) Review on solar adsorption refrigeration cycle. Int J Mech Eng Technol 6340:190–226

46. Anyanwu EE (2004) Review of solid adsorption solar refrigeration II: An overview of the principles and theory. Energy Convers Manage 45:1279–1295

47. Jiang B, Guo H (2014) Prediction of mode specificity, bond selectivity, normal scaling, and surface lattice effects in water dissociative chemisorption on several metal surfaces using the sudden vector projection model. J Phys Chem C 118(46):26851–26858

48. Jiang B, Guo H (2015) Quantum and classical dynamics of water dissociation on Ni(111): A test of the site-averaging model in dissociative chemisorption of polyatomic molecules. J Chem Phys 143:1

49. Jackson B, Nattino F, Kroes GJ (2014) Dissociative chemisorption of methane on metal surfaces: tests of dynamical assumptions using quantum models and ab initio molecular dynamics. J Chem Phys 141:163

50. Liu T, Fu B, Zhang DH (2014) Validity of the site-averaging approximation for modeling the dissociative chemisorption of H_2 on Cu(111) surface: a quantum dynamics study on two potential energy surfaces. J Chem Phys 141:184705
51. Touzain P (1999) Thermodynamic values of ammonia-salts reactions for chemical sorption heat pumps. In: Proceedings of international sorption heat pump conference, Munich, Germany, pp 24–26
52. Neveu P, Castaing J (1993) Solid-gas chemical heat pumps: Field of application and performance of the internal heat of reaction recovery process. Heat Recovery Syst CHP 13:233–251
53. Mbaye M, Aidoun Z, Valkov V, Legault A (1998) Analysis of chemical heat pumps (CHPS): basic concepts and numerical model description. Appl Therm Eng 18:131–146
54. Zhang YH (1989) Adsorption, scientific and technical documents publishing house (in Chinese)
55. Wang LW (2005) Study on chemical adsorption precursor of $CaCl_2$-NH_3 in adsorption refrigeration. Sci China Technol (in Chinese) 35:31–42
56. Wang LW, Wang RZ, Wu JY, Wang K (2004) Adsorption performances and refrigeration application of adsorption working pair of $CaCl_2$-NH_3. Sci China Ser E 47(2):173–185
57. Wang LW, Wang RZ, Wu JY, Wang K (2005) Research on the chemical adsorption precursor state of $CaCl_2$-NH_3 for adsorption refrigeration. Sci China Ser E 48(1):70–82
58. Ci YX, Zhou TZ (1999) The multiple complex compounds in the analytical chemistry (in Chinese, ISBN 7-03-007128-X/O·1069). Sci Press, Beijing, China
59. Gasser RPH (1987) An introduction to chemisorption & catalysis by metals. Clarendon Press, Oxford, UK
60. Biltz W, Huttig GF (1920) Uber die auswertung von dissoziationsmessungen bei ammoniakaten nach dem theorem von nernst mit hilfe von nomogrammen. Z Anorg Allg Chem 109:111–125
61. Xia SW (1993) Activated energy and its calculation (in Chinese, ISBN 7040036312). High Education Press of Beijing, Beijing, China
62. Zhang YH (1989) Adsorption action (in Chinese, ISBN 7805134979, 9787805134970). Shanghai Press of Science and Technology, Shanghai, China
63. Peng SP, Wang B, Luo ZJ (1984) Structure of atoms and molecules, complex compounds, colloid chemistry (in Chinese, ISBN 7118.814). People's Press of Sichuan Province. Chengdu, China
64. Wang LW, Wang RZ, Wu JY, Wang K (2004) Study on adsorption characteristics of calcium chloride ammonia and its application in refrigeration. Sci China Technol (in Chinese) 34:268–279
65. Peng SF (1984) Inorganic chemistry atomic structure complex colloid chemistry. Sichuan People's Publishing House (in Chinese)
66. Zhong Y, Critoph RE, Thorpe RN, Tamainot-Telto Z, Aristov Yu (2007) Isothermal sorption characteristics of the $BaCl_2$-NH_3 pair in a vermiculite host matrix. Appl Therm Eng 27:2455–2462
67. Wang LW (2005) Adsorption characteristics and mechanism of a new type of composite adsorbent and its application in high efficiency heat pipe waste heat refrigeration (in Chinese). Shanghai Jiao Tong University, Shang hai
68. Cengel, YunusA (2002) Thermodynamics: an engineering approach, 4th ed.
69. Trudel J, Hosatte S, Ternan M (1999) Solid-gas equilibrium in chemical heat pumps: the NH_3-$CoCl_2$ system. Appl Therm Eng 19:495–511
70. Aidoun Z, Ternan M (2001) Pseudo-stable transitions and instability in chemical heat pumps: the NH_3-$CoCl_2$ system. Appl Therm Eng 21:1019–1034
71. Zhou ZS, Wang LW, Jiang L et al (2016) Non-equilibrium sorption performances for composite sorbents of chlorides—ammonia working pairs for refrigeration. Int J Refrig 65:60–68
72. Simonova IA, Aristov Yu (2005) Sorption properties of calcium nitrate dispersed in silica gel: the effect of pore size. Russ J Phys Chem 79:1307–1311
73. Wang LW, Wang RZ, Oliveira RG (2009) A review on adsorption working pairs for refrigeration. Renew Sust Energy Rev 13:518–534

74. Critoph RE (1998) Performance limitations of adsorption cycles for solar cooling. Sol Energy 41:21–31
75. Marcio S, Jose CP (2007) Optimum reference temperature for reparameterization of the Arrhenius equation. Part 1: problems involving one kinetic constant. Chem Eng Sci 62:2750–2764
76. Gao J, Wang LW, Wang RZ, Zhou ZS (2017) Solution to the sorption hysteresis by novel compact composite multi-salt sorbents. Appl Therm Eng 111:580–585
77. An GL, Wang LW, Gao J, Wang RZ (2019) Mechanism of hysteresis for composite multi-halide and its superior performance for low grade energy recovery. Sci Reports 9(1):1563
78. Daou K, Wang RZ, Xia ZZ (2006) Development of a new synthesized adsorbent for refrigeration and air conditioning applications. Appl Therm Eng 26:56–65
79. Maggio G, Gordeeva LG, Freni A, Aristov Yu, Santori G, Polonara F et al (2009) Simulation of a solid sorption ice-maker based on the novel composite sorbent lithium chloride in silica gel pores. Appl Therm Eng 29:1714–1720
80. Dawoud B (2007) A hybrid solar-assisted adsorption cooling unit for vaccine storage. Renew Energy 32:947–964
81. Santamaria S, Sapienza A, Frazzica A, Freni A, Girnik IS, Aristov Yu (2014) Water adsorption dynamics on representative pieces of real adsorbers for adsorptive chillers. Appl Energy 134:11–19
82. Sapienza A, Santamaria S, Frazzica A, Freni A, Aristov Yu (2014) Dynamic study of adsorbers by a new gravimetric version of the Large Temperature Jump method. Appl Energy 113:1244–1251
83. Veselovskaya JV, Tokarev MM (2011) Novel ammonia sorbents "porous matrix modified by active salt" for adsorptive heat transformation: 4. Dynamics of quasi-isobaric ammonia sorption and desorption on $BaCl_2$ /vermiculite. Appl Therm Eng 31:566–572
84. Veselovskaya JV, Tokarev MM, Grekova AD, Gordeeva LG (2010) Novel ammonia sorbents porous matrix modified by active salt for adsorptive heat transformation: 6. The ways of adsorption dynamics enhancement. Appl Therm Eng 30:584–589
85. Chakraborty A, Saha BB, Ng KC, Koyama S, Srinivasan K (2009) Theoretical insight of physical adsorption for a single-component adsorbent+adsorbate system: I. Thermodynamic property surfaces. Langmuir Acs J Surf Colloids 25:2204
86. Chakraborty A, Saha BB, Ng KC, Koyama S, Srinivasan K (2009) Theoretical insight of physical adsorption for a single component adsorbent+ adsorbate system: II. Henry Region. Langmuir 25(13):7359–7367

Chapter 5
Solid Sorption Cycle for Refrigeration, Water Production, Eliminating NO$_x$ Emission and Heat Transfer

Abstract Composite sorbents in single-stage refrigeration cycle, two-stage freezing cycle, semi-open solar-driven sorption air-to-water cycle, NH$_3$ sorption cycle, NO$_x$ converting cycle and solid sorption heat pipe cycle for refrigeration, eliminating NO$_x$ emission, water production and heat transfer are summarized in this chapter, which promote the relative commercial utilization in energy conversion fields.

Keywords Sorption cycle · Refrigeration · Freezing · Water production · NO$_x$ emission · Heat transfer

5.1 Solid Sorption Cycle for Refrigeration

There are two conditions for refrigeration: one is for air conditioning with the refrigeration temperature higher than 0 °C, and the other is for freezing requirement with refrigeration temperature below 0 °C. Nowadays, extensive research has been conducted on air conditioning conditions with the refrigeration temperature higher than 0 °C and the heat source temperature lower than 100 °C. Restuccia et al. [1] developed a sorption chiller that uses silica gel impregnated with CaCl$_2$ as the sorbent. The COP is close to 0.6 when the power generation temperature is 85–95 °C and the coolant temperature is 35 °C. Saha et al. [2] studied a three-stage silica gel/water sorption chiller, which can produce cooling power when the heat source temperature is as low as 50 °C and the cooling temperature is 30 °C. Shanghai Jiao Tong University developed a small silica gel water sorption chiller, which was successfully commercialized in 2005 [3]. Kasiwagi [4] proposed a multi-stage cycle that can be driven by waste heat at 50 °C, and it mainly uses silica gel-water as a working pair [5].

But for freezing conditions, the silica gel-water working pair cannot be used. The sorption working pairs, including activated carbon-methanol, activated carbon-ammonia and chlorides-ammonia, which can be driven by low grade heat, have been researched. For example, the research done by Erhard and Hahne [6] and Erhard et al. [7] used the SrCl$_2$-NH$_3$ as working pair. The cooling temperature obtained is as low as −10 °C, but the COP is only about 0.045–0.082, while the heat source

© Science Press 2021

L. Wang et al., *Property and Energy Conversion Technology of Solid Composite Sorbents*, Engineering Materials, https://doi.org/10.1007/978-981-33-6088-4_5

temperature needs to be as high as 100–120 °C. Pons and Guilleminot [8] developed a prototype using an activated carbon-methanol pair, and solar panels can produce nearly 6 kg of ice per m^2 with the amount of sunshine per day of about 20 MJ.

For the freezing temperature lower than −15 °C, the heat source temperature needs to be higher than 100 °C. For example, Wang et al. [9] developed a split-flow heat pipe sorption ice maker with the solid sorbent of $CaCl_2$ impregnated with activated carbon. The COP and SCP are 0.41 and 731 W/kg, respectively, and the heat source temperature is 110–120 °C. The multi-stage cycle can reduce the heat source temperature requirement for sorption systems. Meunier [10] analyzed the heat integration problem by introducing cascade cycles. Aiming at a lower temperature suitable for deep-freezing while using a lower heating source temperature, a cascaded thermochemical system adopting $BaCl_2$ salt reacting with ammonia was developed by Pierres et al. [11]. The system consists of two reactors, two condensers and two evaporators (one evaporator is designed inside one reactor). When the ambient temperature is 25 °C, the evaporation temperature of the system is as low as −33 °C. The disadvantage of this system is its complex structure. Two reactors are used in the cascade ice making system, and the refrigeration effect of one reactor (reactor 1) is thermally linked to the other reactor (reactor 2). The COP of a single system in the cascade system generally is less than 0.4, which means that reactor 1, generating cooling power for cooling reactor 2 in order to reduce the evaporation temperature, should be more than 2 times larger than reactor 2. Such a structure leads to low COP and big volume of the system [11].

To better understand the applications in refrigeration, in this section, several single-stage and two-stage solid sorption refrigeration cycles used for ice making condition will be analyzed in detail.

5.1.1 Single-Stage Solid Sorption Refrigeration Cycle

Conventional solid sorption refrigeration systems are mainly heated by hot oil or water, which have higher heat transfer coefficient than that of gas [12, 13]. For the heat recovery of the exhaust gas, the direct heat transfer process between exhaust gas and sorption bed will make the system compact for that the intermediate heat exchanger between gas and hot oil or water won't be needed [14]. For example, in a two-stage indirect exhaust heat recovery system (recommended for automotive air conditioning), two parallel flow shell and tube heat exchangers are used for heat transfer [15]. A sorption cooling system with one or two sorption beds is established to study the effect of the number of sorption beds on the cooling performance when silicone oil is used as a heating fluid [16]. For such a design an additional heat exchanger is used to transfer heat between exhaust gas and silicone oil. Meanwhile a heat exchanger between sea water and water is required. The system will be simpler if the system can directly exchange heat with exhaust gas and air efficiently. Zhong et al. studied a sorption air conditioning system, for which each module contains a sorbent bed, a heat pipe heat exchanger and an insulating part. The system is heated by exhaust gas and cooled by air [17] directly. However, the authors only conducted simulations and proved the feasibility theoretically, and the performance isn't verified by experiments.

In recent years, the composite solid sorbents [18, 19] with high thermal conductivity and reasonable permeability have been developed [20, 21], and such a progress could improve the heat transfer performance for recovering the thermal energy from exhaust gas directly. The multi-halides were proved could decrease the hysteresis phenomena effectively, i.e. the desorption temperature is lower than single-salt sorbent. It makes the single-stage refrigeration cycle feasible for low-grade heat source temperature and high environmental temperature [22, 23].

5.1.1.1 Working Principle of the Single-Stage Chemisorption Refrigeration System

The working principle of the chemisorption refrigeration cycle, which consists of a sorption bed, a condenser, an evaporator and a liquid storage tank, is shown in Fig. 5.1a, b. A complete cycle includes two phases:

(1) Desorption phase: At this stage, the sorption bed is heated. When the temperature rises to T_d (Fig. 5.1a), ammonia will be desorbed from the sorption bed. The desorbed ammonia flows into the condenser (curve DC in Fig. 5.1a and process a in Fig. 5.1b) and then condenses. Liquid ammonia is stored in a liquid storage tank (process b in Fig. 5.1b and ammonia is shown as point C in Fig. 5.1a).

(2) Sorption phase: At this stage, the sorption bed is cooled. The pressure of liquid ammonia drops sharply after passing through the expansion valve, and then flows into the evaporator (curve CB in Fig. 5.1a and process c in Fig. 5.1b). The supply air is cooled by the evaporation of ammonia in the evaporator, and the

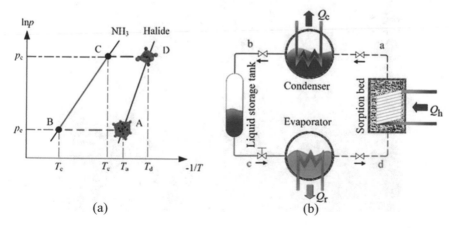

(a)   (b)

Fig. 5.1 Working principle of the single-stage chemisorption refrigeration cycle. **a** Chemisorption principle; **b** sorption refrigeration cycle [24]

cold gaseous ammonia flows back to the sorption bed (curve BA in Fig. 5.1a and process d in Fig. 5.1b), and finally completes the cycle.

Such a cycle is applied to a refrigerated truck used as a refrigerator, as shown by the arrow in Fig. 5.2a, with the sorption bed installed on one side of the girder. Compared with conventional compression refrigeration systems, the position of the condenser remains unchanged. The evaporator is installed in the storage room of the truck. The liquid storage tank is installed under the condenser. The system is directly driven by the exhaust gas of the engine. In the test unit, an electric heater (as shown in Fig. 5.2b) is designed to heat the circulating air to simulate the exhaust gas. The gas circulation heating system includes an electric heater for generating hot air, a high temperature fan for hot air circulation, and some additional pipes for rapidly increasing the temperature of the sorption bed during the desorption process. As shown in Fig. 5.2a, the hot air temperature at the outlet of the electric heater is

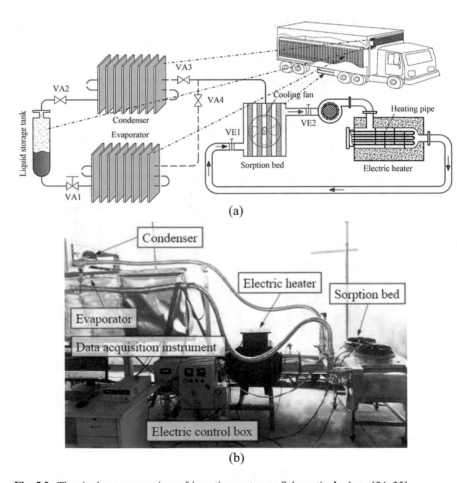

Fig. 5.2 The single-stage sorption refrigeration system. **a** Schematic; **b** photo [24, 25]

monitored and controlled by the feedback system. The large box on the middle of Fig. 5.2b is a simulated freight compartment in which a small electric heater is used to balance the cooling power.

The test procedures can be described as follows:

(1) Open the exhaust gas valve (VE1, VE2) and start the electric heater. The sorption bed is heated by the hot air and the internal pressure will increase. The ammonia valve between the sorption bed and the condenser (VA3) remains closed until the pressure in the sorption bed increases above the pressure in the condenser. The condenser fans and valve (VA2) are then opened while valves VA1 and VA4 remain closed during the desorption process. When the desorption process is finished, the electric heater and the condenser fans stop, and then close valves including VE1, VE2, VA2 and VA3.

(2) Open the cooling air valve and turn on the cooling fans. The sorption bed is then cooled by the ambient air and the pressure inside it will decrease. The ammonia valve between the evaporator and the sorption bed (VA4) keeps closed until the pressure inside the sorption bed decreases below the pressure in the evaporator. Then turn on the small electric heater in the refrigerated compartment and the evaporator fans. The expansion valve (VA1) is adjusted for a desired evaporating pressure. Valves VA2, VA3, VE1 and VE2 keep closed during the sorption process. When the sorption process is completed, stop the small electric heater, cooling fans and evaporator fans and then close the previously opened valves.

5.1.1.2 Design of Sorption Bed Heating/cooling by Exhaust Gas/air

Due to the high thermal diffusivity of the ENG-TSA and its reduced sorption hysteresis, the multi-salt sorbent of $CaCl_2/MnCl_2$ with ENG-TSA as matrix is proved to be superior to the existing sorbents for gas cooling or heating process. A reasonable design of the sorption bed is another key element. First of all, in view of the hot summer weather and the small heat transfer coefficient of the gas, the heat transfer area should be expanded, which is contrary to the miniaturization of the sorption bed. Secondly, the heat transfer coefficient of air is smaller than that of exhaust gas. For the two heat transfer fluids, the heat transfer enhancement methods should be different. In addition, the temperature difference between the exhaust gas and the sorption bed is much larger than the temperature difference between the cooling air and the sorption bed, which results in different flow rates. The final design scheme is shown in Fig. 5.3.

The engine exhaust gas flows perpendicular to the direction of the cooling air at high temperature and low flow rate. Generally, the ambient temperature in summer exceeds 30 °C and a larger air flow rate is required to cool the sorption bed. In that case, only three layers of unit tubes are arranged in the cooling direction to maintain a better cooling effect. Moreover, the axial flow fans with big air volume and low power consumption are installed to achieve a rapid cooling. The staggered tube bundles make the heat exchange area larger and can also interfere the flow of the exhaust gas and cooling air better.

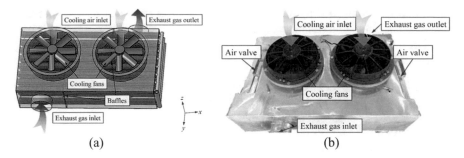

(a)   (b)

Fig. 5.3 Structure of the gas heating/cooling sorption bed. **a** Schematic; **b** photo [24]

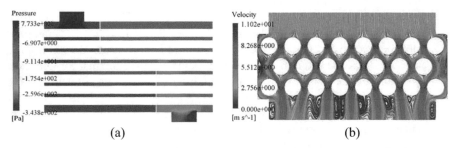

(a)   (b)

Fig. 5.4 Flow characteristics of the sorption bed. **a** Pressure drop of the exhaust gas; **b** flow behavior of the cooling air [24]

For the exhaust gas recovered from the engine, an important factor is the pressure drop of the exhaust gas between the inlet and the outlet of the sorption bed. It cannot be too high, otherwise it will affect the dynamic characteristics of the vehicle. The numerical simulation results of the sorption bed are shown in Fig. 5.4.

As shown in Fig. 5.4a, when the air flow rate is 600 m³/h, the pressure decreases in the flow direction, and the pressure difference between the inlet and the outlet of the sorption bed is approximately 380 Pa. This data is much smaller than the exhaust back pressure (3 kPa) of turbo-charged diesel engines, so it will not have a significant impact on engine performance. The path of the cooling air is shown in Fig. 5.4b. According to the rated flow of the cooling fan, the inlet speed is assumed to be 3 m/s. The flow velocity near the tube wall can even reach a maximum of about 10 m/s, while the flow velocity near the bottom of the last row of unit tubes (especially four of them) reaches a minimum. Under this condition, the sorption bed can still be uniformly cooled by cooling air.

5.1.1.3 Thermodynamic Analysis

The heat transferred from/to the sorption bed is calculated by the temperature difference between the inlet and outlet of the heating/cooling air. The refrigeration capacity is calculated similarly by the temperature difference between the supply air and return air, given by the following expression:

$$q_i = \rho_i V_i c_i \left(T_{h(i)} - T_{l(i)} \right) / 3600 \tag{5.1}$$

where q_i is the heat transfer rate (kW; i = r: for refrigeration phase, q_r is the refrigeration capacity); ρ_i is the air density (kg/m^3); i = c: for cooling phase, q_c is the heat carried away by the cooling air during sorption process; i = h: for heating phase, q_h is the heat transferred to the sorption bed during desorption process; V_i is the volumetric flow rate of air (m^3/h); c_i is the specific heat of air (kJ/(kg K)); $T_{h(i)}$ and $T_{l(i)}$ are the relatively high and low temperature of the inlet and outlet temperature respectively (°C) (i = h, hot air; i = c, cooling air; i = r, chilled air for refrigeration).

The COP, ignoring the heat released to the environment by pipes and air leakage, and the SCP are then calculated by:

$$COP = \frac{q_r}{q_h + w_e/\eta} \tag{5.2}$$

$$SCP = q_r/m_a \tag{5.3}$$

where w_e is the total rated power of the condenser fans, evaporator fans and cooling fans (kW); m_a is the mass of the multi-salt sorbent (kg); the thermoelectric conversion efficiency η is taken as 0.35.

5.1.1.4 Performance Analysis of the Single-Stage Sorption Refrigeration System

The exhaust gas temperature changes with the engine load, thus it is necessary to study the influence of the heating temperature on the refrigeration system. The summer average temperature of most cities in southern China is 25–30 °C. In order to test the refrigeration performance under this extreme conditions, such as low heating temperature and high ambient temperature, the preset heating temperature range is 200–250 °C and the ambient temperature is about 25–30 °C.

Three key performance parameters (the average refrigeration capacity q_r, COP and SCP) are calculated to evaluate the refrigeration performance at different heating temperatures. The results are given in Fig. 5.5. The supply air temperature of the evaporator is controlled at −5 °C and the cycle time is 65 min (30 min for sorption, 30 min for desorption and 5 min for transition).

As shown in Fig. 5.5a, when the heating temperature changes from 200 to 250 °C, the average refrigeration capacity of the entire cycle increases by less than 10%. The results show that when the engine load changes, the sorption refrigeration system can stably generate cooling power for refrigerated truck. Under low-load engine operating conditions (i.e., the exhaust gas temperature is only 200 °C), the average

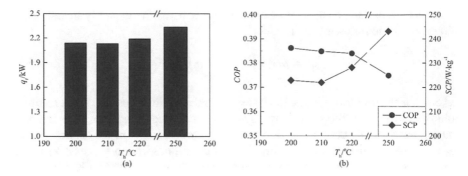

Fig. 5.5 The refrigeration performance versus heating temperature. **a** Refrigeration capacity; **b** COP&SCP [25]

cooling capacity is about 2.14 kW, which is enough to meet the cooling needs of light-duty refrigerated vehicles. The SCP in Fig. 5.5b shows a similar trend in terms of average cooling capacity. Ignoring experimental error, SCP increases with heating temperature. When the heating gas is 250 °C, the maximum SCP of the sorption refrigeration system is greater than 240 W/kg. At the same time, COP is related to both cold output and heat input, so the range of COP change is small. The heat input to the system increases with the increase of the heating temperature. When the heating temperature is 200–250 °C, the COP is maintained at about 0.38.

The required refrigeration capacity is strongly linked to the refrigeration temperature in most cases. Thus the characteristics of the sorption freezing system at different refrigeration temperatures are analyzed.

Figure 5.6a shows that when the refrigeration temperature changes from −15 to 0 °C and then flattens out, the average refrigeration capacity of the entire cycle increases almost linearly. The lower system pressure means that the refrigeration temperature is lower during the sorption process, and the sorbent inside the sorption bed needs to be cooled to a lower temperature to perform the synthesis

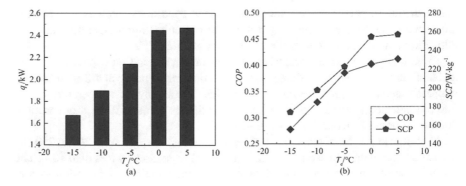

Fig. 5.6 The refrigeration performance versus refrigeration temperature. **a** Refrigeration capacity; **b** COP&SCP [25]

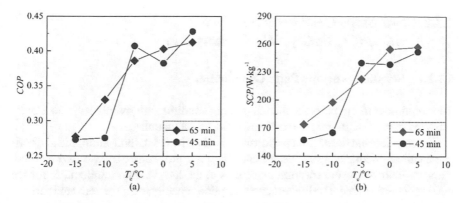

Fig. 5.7 The refrigeration performance versus cycle time. **a** COP; **b** SCP [25]

reaction at a lower pressure. As the refrigeration temperature decreases, the sorption rate and refrigeration capacity both decrease. The largest refrigeration capacity is about 2.5 kW at 5 °C refrigeration temperature, and the corresponding maximum COP and SCP shown in Fig. 5.6b are approximately 0.41 and 260 W/kg respectively. The refrigeration capacity is about 1.67 kW when the refrigeration temperature is −15 °C, which can still meet the demand of the light truck for refrigerated transportation. As shown in Fig. 5.6b, COP and SCP both increase with the refrigeration temperature. The COP increases slightly faster than the SCP when the refrigeration temperature range is from −15 to −5 °C, but after that, the COP increases slower.

Cycle time is another important factor which affects the refrigeration performance. The characteristics of the sorption refrigeration system are investigated and compared under the cycle time of 65 min and 45 min (20 min for desorption, 20 min for sorption and 5 min for transition).

Figure 5.7a shows that under different cycle times, COP increases as the cooling temperature increases. The COP for a 65 min period is higher than the COP for a 45 min period. The COP of the 65 min cycle continues to increase, while the COP of the 45 min cycle remains around 0.28 when the cooling temperature is below −10 °C, and then fluctuates around 0.4 when the refrigeration temperature changes from −5 to 5 °C. The sorption bed is directly cooled by the ambient air, and the ambient temperature will change during the day. The ambient temperature of the 65 min cycle at −5 °C cooling temperature is about 5 °C higher than the ambient temperature of the 45 min cycle, so the COP of the 65 min cycle is lower. The SCP and COP in Fig. 5.7b show similar trends. When the refrigeration temperature changes from −5 to 5 °C, the cycle of 65 min has no obvious advantages, but the facts have proved that it is superior in the entire refrigeration temperature range. The minimum SCP for a 45 min cycle is about 158 W/kg, and the corresponding cooling capacity is 1.5 kW at a cooling temperature of −15 °C. It can be concluded from Fig. 5.7 that for the sorption refrigeration system, the time distribution of the 65 min cycle is more reasonable than the time distribution of the 45 min cycle.

5.1.2 Two-Stage Chemisorption Cycle
with CaCl₂/BaCl₂-NH₃ Working Pair

5.1.2.1 Sorption Working Pairs Comparison

The sorption working pairs for the ice making condition are mainly salts-ammonia, activated carbon-ammonia, and activated carbon-methanol.

The sorption and desorption performances of activated carbon-ammonia and activated carbon-methanol were tested, and the main problems of these two working pairs are their low cycle sorption capacities at the low heating temperature for ice making condition [26]. The highest cycle sorption capacity in that of activated carbon-ammonia and activated carbon-methanol is only 0.02 kg/kg while the evaporating temperature, environmental temperature and heating temperature are −20, 30, and 100 °C respectively. Moreover, the cycle sorption capacity is almost 0 when the desorption temperature is lower than 90 °C.

The salt-ammonia sorption working pairs that could be possibly utilized for the low temperature heat source are mainly $BaCl_2$-NH_3, $SrCl_2$-NH_3 and $CaCl_2$-NH_3.

The problem of $BaCl_2$-NH_3 working pair is that the cooling temperature for the sorbent needed in the freezing working condition is too low, as shown in the Fig. 5.8 [27]. The $BaCl_2$-ammonia working pair could desorb at a temperature of 71 °C (point B) while the condensing temperature is 35 °C (point A). However, for the evaporating temperature of −25 °C (point D), the cooling temperature for the sorbent has to be lower than 21 °C (point E). This implies the $BaCl_2$-NH_3 working pair cannot be utilized for an evaporating temperature of −25 °C in summer, because the cooling temperature is normally higher than 32 °C.

The common drawback of the $SrCl_2$-NH_3 and $CaCl_2$-NH_3 working pairs is their high desorption temperature. As shown in Fig. 5.8, a freezing temperature of −25 °C (point D) will be obtained while the cooling temperature of $SrCl_2$ and $CaCl_2$ is lower than 50 °C (point F). This is easy to be obtained even in summer. However, for the

Fig. 5.8 The Clapeyron diagram of a chemisorption cycle [27]

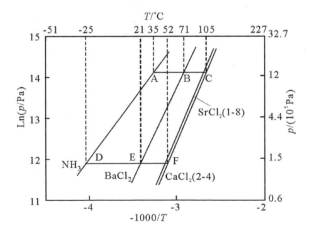

desorption process, at a condensing temperature of 35 °C (point A), the desorption temperature must be higher than 100 °C (point C).

5.1.2.2 Design of the Two-Stage Chemisorption Cycle

In a conventional sorption or resorption cycle, the operation process is usually carried out at two pressure levels, namely, high pressure in the regeneration stage of the reaction salt and low pressure in the cold production stage. For such a cycle, when freezing output is required, the required heat source temperature is too high or the required coolant temperature of the solid sorbent is too low. In order to complete the refrigeration under freezing conditions at low heat source temperature and reasonable coolant temperature, a two-stage sorption cycle is proposed by combining the conventional sorption cycle and resorption cycle [28].

A two-stage deep freezing cycle consists of two types of sorbents, i.e. $BaCl_2$ and $CaCl_2$. The operation of the cycle involves a two-stage process as shown in Fig. 5.9, in which the sorbent of $BaCl_2$ serves as the low temperature sorbent, while the sorbent of $CaCl_2$ serves as the high temperature sorbent.

During the first stage, the high temperature sorbent is heated by a heat source Q_{H1}, while the low temperature sorbent is cooled by a cooling source Q_{L2} at ambient temperature. The high temperature sorbent is linked with the low temperature sorbent, and the refrigerant desorbed from the high temperature sorbent is sorbed by the low temperature sorbent. In the second stage, the low temperature sorbent is heated by the heat source Q_{H2}, and the condenser is cooled by the cooling source at ambient temperature. The low temperature sorbent is linked with condenser, and then the refrigerant desorbed from the low temperature sorbent condenses in the condenser. The refrigeration process also proceeds during the second stage. The high temperature sorbent is cooled by the cooling source Q_{L1} at the ambient temperature. The high temperature sorbent is linked with evaporator. The refrigerant inside the evaporator evaporates and is sorbed by the high temperature sorbent, meanwhile this process generates cooling power.

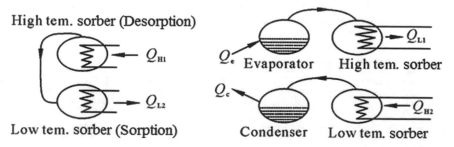

Fig. 5.9 Working processes of two-stage deep freezing cycle. **a** The first stage; **b** the second stage [28]

Fig. 5.10 The Clapeyron diagram for the two-stage chemisorption cycle [27]

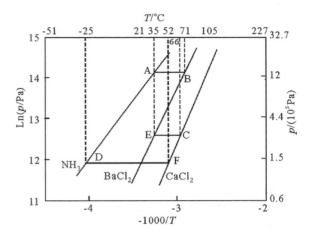

The Clapeyron diagram for this new chemisorption cycle is shown in Fig. 5.10 [27]. During the first desorption stage, supposing the $BaCl_2$ sorbent is cooled by the cooling water with a temperature of 35 °C, the synthesis point of ammonia and $BaCl_2$ is at E, then the decomposing point of ammoniate calcium chloride is point C, which corresponds to the desorption temperature of 66 °C. In the second stage, while the condensing temperature is 35 °C, the desorbing point of ammoniate barium chloride is point B, which corresponds to the desorbing temperature of 71 °C. For the sorption process between evaporator and $CaCl_2$ sorbent, while the evaporating temperature is −25 °C, the cooling temperature needed for $CaCl_2$ is lower than 52 °C, which is easy to be obtained while the temperature of cooling water is 35 °C. For such a cycle the theoretical desorption temperature is lower than 80 °C while the freezing temperature of −25 °C could be obtained, then it could be utilized for the solar energy utilization and the industrial waste heat recovery at the temperature between 70 and 80 °C.

5.1.2.3 Sorption Performance Test of the Sorbents

In order to test performances of $CaCl_2$ and $BaCl_2$, the composite sorbents which utilize expanded graphite as a binder are produced and tested, and the characteristic parameters of sorbents are shown in Table 5.1.

Table 5.1 Characteristic parameters of sorbents [28]

Composite sorbent	$CaCl_2$/Expanded graphite	$BaCl_2$/Expanded graphite
Mass (g)	133.18/98.39	187.41/44.16
Density (kg/m^3)	400	400
Size: outlet diameter × inlet diameter × length (mm)	51 × 12 × 300	51 × 12 × 300

Fig. 5.11 The sorption performance test unit [28]

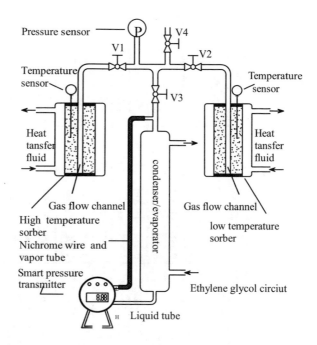

In order to test the sorption and desorption performances of $CaCl_2$ and $BaCl_2$, the test unit in Fig. 5.11 is designed. The test unit mainly includes three parts, which are high temperature sorbent reactor, low temperature sorbent reactor, and one condenser/evaporator. $CaCl_2$ is put in the high temperature sorbent reactor, and $BaCl_2$ is put in the low temperature sorbent reactor. Both sorbents are heated and cooled by water in the experiments. The temperature of the refrigerant of ammonia in the condenser/evaporator is controlled by the ethylene glycol circuit linked with cryostat. The sorption capacities of sorbents are mainly calculated by the change of pressure difference between the vapor tube linked with the upper vapor end of ammonia and the liquid tube linked with bottom liquid end of ammonia of the condenser/evaporator, which is tested by a smart pressure transmitter. A nichrome wire is winded at the vapor tube in order to prevent the condensation of the ammonia vapor.

The experimental results are shown in Tables 5.2 and 5.3, which are sorption quantities of $CaCl_2$ after sorption/$BaCl_2$ after desorption, and the cycle sorption capacities of $CaCl_2$/$BaCl_2$, under the heating, cooling and evaporating temperatures of 70–90 °C, 25–35 °C and −25 to 5 °C, respectively.

5.1.2.4 Design and Construction of Sorption Freezing Machine

A two-stage sorption freezing machine is designed [29] and the diagram is shown in Fig. 5.12. One $CaCl_2$ sorbent bed, i.e. middle temperature sorption bed (MTS bed), $BaCl_2$ sorbent bed, i.e. low temperature sorption bed (LTS bed), a condenser and

Table 5.2 Sorption quantities of CaCl₂ after sorption and BaCl₂ after desorption [28]

Temperature of heat source (°C)	Cooling/condensing temperature T_c/°C	Sorption quantity of CaC₂ after sorption/kg/kg							Sorption quantity of BaCl₂ after desorption/kg/kg
		Evaporating temperature T_e/°C							
		−25	−20	−15	−10	−5	0	5	
70	25	0.356	0.394	0.443	0.484	0.538	0.598	0.669	0.419
	30	0.256	0.302	0.371	0.414	0.459	0.508	0.557	0.400
	35	0.121	0.136	0.198	0.233	0.272	0.316	0.371	0.364
75	25	0.352	0.435	0.474	0.531	0.575	0.619	0.701	0.462
	30	0.286	0.368	0.413	0.456	0.507	0.563	0.638	0.439
	35	0.269	0.328	0.358	0.400	0.448	0.497	0.541	0.414
80	25	0.425	0.536	0.584	0.627	0.683	0.745	0.819	0.500
	30	0.402	0.490	0.514	0.563	0.604	0.655	0.714	0.475
	35	0.321	0.400	0.453	0.499	0.550	0.607	0.656	0.446
85	25	0.556	0.598	0.642	0.690	0.743	0.791	0.865	0.528
	30	0.422	0.489	0.495	0.543	0.590	0.643	0.700	0.505
	35	0.357	0.404	0.433	0.479	0.524	0.567	0.611	0.482
90	25	0.515	0.560	0.647	0.675	0.749	0.800	0.870	0.573
	30	0.514	0.552	0.600	0.646	0.694	0.738	0.800	0.549
	35	0.397	0.419	0.454	0.493	0.538	0.591	0.638	0.519

Table 5.3 Cycle sorption quantities of the system [28]

Temperature of heat source (°C)	Cooling/condensing temperature T_c/°C	Evaporating temperature T_e/°C						
		−25	−20	−15	−10	−5	0	5
70	25	–	–	0.024	0.065	0.538	0.598	0.669
	30	0.256	0.302	0.371	0.414	0.459	0.508	0.557
	35	0.121	0.136	0.198	0.233	0.272	0.316	0.371
75	25	0.352	0.435	0.474	0.531	0.575	0.619	0.701
	30	0.286	0.368	0.413	0.456	0.507	0.563	0.638
	35	0.269	0.328	0.358	0.400	0.448	0.497	0.541
80	25	0.425	0.536	0.584	0.627	0.683	0.745	0.819
	30	0.402	0.490	0.514	0.563	0.604	0.655	0.714
	35	0.321	0.400	0.453	0.499	0.550	0.607	0.656
85	25	0.556	0.598	0.642	0.690	0.743	0.791	0.865
	30	0.422	0.489	0.495	0.543	0.590	0.643	0.700
	35	0.357	0.404	0.433	0.479	0.524	0.567	0.611
90	25	0.515	0.560	0.647	0.675	0.749	0.800	0.870
	30	0.514	0.552	0.600	0.646	0.694	0.738	0.800
	35	0.397	0.419	0.454	0.493	0.538	0.591	0.638

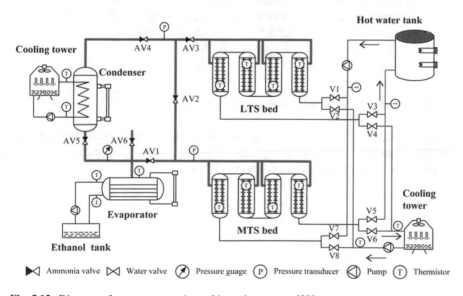

Fig. 5.12 Diagram of two-stage sorption refrigeration system [29]

an evaporator are included in this system. Both the MTS bed and the LTS bed are designed by four unit reactors. For high-pressure systems, this design can reduce the thickness for the wall of steel vessel, which can reduce the ratio for the heat capacity of the steel if compared with that for sorbent. Each reactor is composed of 7 finned tubes, which are made of steel tubes and aluminum fins. A composite sorbent made by chloride and expanded natural graphite is filled between fins. The mass ratio between chloride and expanded natural graphite is 4:1, and the density of the composite sorbent is about 300 kg/m^3.

The working processes of the system are as follows:

(1) Resorption process. During this process, the water valves V2, V4, V5, V7 are opened, and the hot water produced by the hot water boiler is used to simulate low grade heat and provide the decomposition heat of MTS bed, and the synthesis heat of the LTS bed is removed by the cooling water in the cooling tower. At the same time, the ammonia decomposed from the MTS bed passes through ammonia valves AV2 and AV3, and then enters the LTS bed, where the synthesis process between ammonia and LTS occurs.

(2) Desorption/sorption process. During this process, the water valves V1, V3, V6, V8, and the ammonia valves AV1, AV3 and AV4 are opened. The LTS bed is heated by hot water, and the ammonia inside the LTS bed is desorbed and then condensed in the condenser. At the same time, the MTS bed is cooled by cooling water and sorbs ammonia from the evaporator. The evaporation process of ammonia inside the evaporator provides a cooling effect. The ethanol aqueous solution circuit is used to transport the cooling quantity out of the evaporator. A photo of the system is shown in Fig. 5.13. The size of the entire test unit (Fig. 5.13a) is 1000 mm × 600 mm × 1500 mm.

5.1.2.5 Performance Evaluation and Test Results

The performance of the system is evaluated by 4 parameters, which are Δx (cycle sorption capacity), W_{ref} (cooling capacity), COP, and SCP.

The cycle sorption capacity of ammonia Δx is:

$$\Delta x = m_{NH_3} = \rho \nu_{NH_3} \qquad (5.4)$$

where ρ is the density of the saturated liquid ammonia, and ν_{NH3} is the volume of the saturated liquid ammonia.

The cooling capacity is:

$$W_{ref} = \frac{Q_e - Q_{re}}{t} = \frac{\int_0^t \left[\dot{m}_{eth} C_{eth} (T_{e,in} - T_{e,out}) \right] - m_{ca} \Delta x C_{NH_3} (T_c - T_e)}{t} \qquad (5.5)$$

where Q_e is cooling amount generated in the evaporator and removed by heat transfer fluid, which is ethanol. Q_{re} is cooling amount consumed when the refrigerant flows

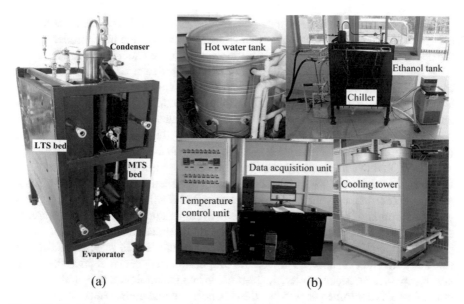

Fig. 5.13 Photograph of two-stage sorption refrigeration test unit. **a** Chiller, **b** test system [29]

from condenser to evaporator. t is cycle time. $\dot{m}_{eth}$ is mass flow rate of ethanol solution. $T_{e,in}$ and $T_{e,out}$ are the temperature of evaporator inlet and outlet, respectively. m_{ca} is the mass of $CaCl_2$ composite sorbent. C_{NH3} is the specific heat of ammonia.

The coefficient of performance is:

$$COP = W_{ref}/W_h \tag{5.6}$$

where W_h is the heating power and is calculated by:

$$W_h = \frac{Q_h}{t} = \frac{\int_0^t \dot{m}_w C_w (T_{hw,in} - T_{hw,out})}{t} \tag{5.7}$$

where Q_h is heating quantity provided by hot water. $\dot{m}_w$ is the mass flow rate of hot water. C_w is the specific heat of water.

The specific cooling power per kilogram sorbent can be expressed as:

$$SCP = W_{ref}/m_{tot} \tag{5.8}$$

where m_{tot} is total mass of composite sorbent, i.e. both the mass of chlorides and expanded natural graphite are considered.

Considering that the system state is unstable in the first few cycles, the system data is collected under steady-state conditions. In this case, the inlet temperature of the cooling sink, heating source and chilling medium does not change much.

When the temperature of the chilling liquid is controlled at the lowest temperature of about −15 °C, the temperature of the sorbent bed, cooling source and heat source is shown in Fig. 5.14. In Fig. 5.14, the temperature of the hot water inlet and the cooling water inlet marked with green and purple lines are respectively controlled at 85 and 25 °C. For the LTS and MTS in Fig. 5.14, the desorption process occurs on the rising curve, and the sorption process occurs on the falling curve. Figure 5.14 also shows that when the red curve rises, the blue curve always falls, and vice versa, which means that the LTS sorbent and the MTS sorbent always work in different modes. In addition, only the sorption process of MTS will produce a cooling effect, and the sorption of LTS will cause the driving temperature of the MTS to decrease, so the system is used as an intermittent refrigeration process.

For refrigeration output, the ethanol solution in the tank is selected as the heat transfer fluid that exchanges heat with the evaporator. In the ethanol tank, the refrigerating effect is balanced by electrical heater. When the minimum temperature of the ethanol tank is controlled at −15 °C, the ethanol inlet and outlet temperatures of the evaporator are shown in Fig. 5.15. Figure 5.15 shows that at the beginning of refrigeration, due to the high sorption rate, the evaporator outlet temperature drops sharply. As the reaction progresses, the sorption weakens and the heat inside the ethanol tank cannot be balanced, so the outlet temperature gradually rises. Figure 5.15 also shows that when the lowest temperature in the ethanol tank is controlled at −15 °C, the lowest outlet temperature of the sorbent is as low as −18 °C between the evaporator outlet and the inlet due to temperature fluctuations that occur during the sorption refrigeration process. The maximum temperature difference is about 7 °C, which shows that the system has huge refrigeration potential.

Under the above conditions, the pressure of the two sorption beds is shown in Fig. 5.16. The results show that the cycle can be divided into two stages, for which from 0 to 1020 s CaCl$_2$ sorbs refrigerant from the evaporator as well as that BaCl$_2$

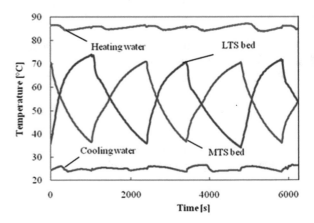

Fig. 5.14 Temperature of sorption beds, heat source and cooling source [29]

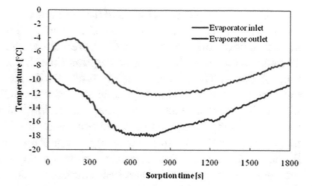

Fig. 5.15 Evaporating temperature under the condition of 85 °C regeneration temperature and 25 °C coolant temperature [29]

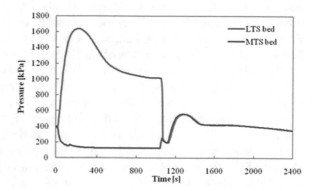

Fig. 5.16 Pressure of two sorbent beds [29]

desorbs to the condenser, and from 1020 to 2400 s the resorption happens between $CaCl_2$ and $BaCl_2$ sorbents. For MTS, because only resorption and sorption occur, its pressure is at a lower value of 150–550 kPa, while for LTS, due to the desorption between the sorbent and the condenser, the pressure is much higher at 230–1620 kPa. The experimental results in Fig. 5.16 illustrate the theoretical analysis of working pressure well, i.e. there are three levels of working pressure, namely 1020, 350 and 130 kPa. It can also be found that there is a peak at the beginning of each stage, which is caused by the high sorption rate.

The second stage is the desorption between the $BaCl_2$ bed and the condenser and the sorption between the $CaCl_2$ bed and the evaporator. Experiments show that the sorption process of $CaCl_2$ is much longer than the desorption process of $BaCl_2$. The reason is mainly because the pressure levels are different in two stages. The sorption of $CaCl_2$ occurs under low pressure, while the desorption of $BaCl_2$ occurs under high pressure. Therefore, the mass transfer performance of the desorption process of $BaCl_2$ is better than that of the sorption process of $CaCl_2$.

The experimental results also illustrate that the sorption of $CaCl_2$ only can be sufficient when the sorption time is controlled at 13 min. Under this sorption time the optimal resorption time between $CaCl_2$ and $BaCl_2$ is studied by the evaluation of COP and SCP at 85 °C regeneration temperature, 25 °C coolant temperature, and

−15 °C evaporating temperature, and the results are shown in Fig. 5.17. In Fig. 5.17 the resorption time varies from 10 to 22 min, and the time gradient is 3 min. It can be seen that the highest SCP and COP are obtained for resorption time of 16 and 19 min, which means the optimal resorption time should be longer than the sorption time of $CaCl_2$. The reason is mainly related with the driving temperature potential. According to the thermodynamic analysis of $CaCl_2$-$BaCl_2$-NH_3, the sorption kinetics is related with the driving temperature lift, which is the difference between the sorbent temperature and the equilibrium sorption/desorption temperature. It is mainly because that the sorption/desorption heat released/absorbed under the equilibrium temperature will be transferred more quickly when the difference between the equilibrium temperature and heat source/heat sink temperature is larger. The driving temperature lifts of first stage desorption, second stage desorption and $CaCl_2$ sorption are 16 °C, 30 °C and 18 °C, respectively, i.e. for the resorption process the driving temperature lift is 16 °C, for the desorption process of $BaCl_2$ the temperature lift is 30 °C, and for the sorption process of $CaCl_2$ the temperature lift is 18 °C. Subsequently, the sorption process of $CaCl_2$ goes faster than resorption process of $CaCl_2$, and the desorption process of $BaCl_2$ is the fastest.

Because the optimal resorption time should be longer than sorption time of $CaCl_2$, several sets of the cycle time with shorter sorption time and longer resorption time of $CaCl_2$ are planned (Table 5.4) for the experiments, for which the sorption time of $CaCl_2$ generally 1/3 to 1/4 less than resorption time, similarly with the ratio between the optimal sorption time for $CaCl_2$ and the resorption time mentioned above. The COP and SCP are studied for different cycle times in Table 5.4, and the heat source temperature, coolant temperature, and refrigerating temperature were 85 °C, 25 °C, and −15 °C in the experiments, respectively. Figure 5.18 shows that SCP increases with the cycle time at the beginning when the cycle time is shorter than 25 min,

Fig. 5.17 SCP and COP versus $CaCl_2$ sorption/desorption time [29]

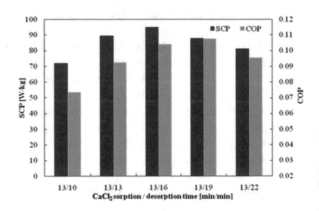

Table 5.4 $CaCl_2$ sorption/desorption time versus cycle time [29]

Cycle time/min	15	20	25	30	35	40	50	60
Sorption/Desorption time/min	6/9	8/12	11/14	13/17	15/20	17/23	22/28	27/33

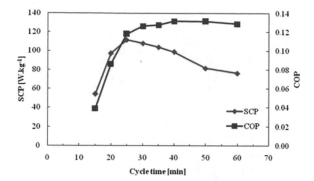

Fig. 5.18 SCP and COP versus cycle time [29]

and then it decreases rapidly with the increasing cycle time when it is longer than 25 min. COP increases with the increasing cycle time and the maximum value is obtained when the cycle time is 25–30 min. Then it almost maintains at a constant value and decreases very slowly. It is mainly because of that both the SCP and COP are the functions of refrigeration capacity. At the beginning the refrigeration capacity increases significantly with the increasing cycle time because more saturated state of sorption and more thorough state of desorption can be gotten with the increasing cycle time. The sorption time and resorption time fit each other best, when the cycle time is 25 min. If the cycle time is prolonged further the refrigeration capacity decreases, thus the SCP decreases. For COP, it is also relative with the heating power and decreases with the cycle time as shown in Fig. 5.19, thus it almost maintains at a constant value when the cycle time is longer than 25 min. Figure 5.18 indicates that the highest SCP is 117.9 W/kg when the cycle time is 25 min while COP reaches its peak 0.13 at 40 min. When the cycle time is 25–40 min the cooling performance of the chiller keeps relatively steadily, and SCP and COP are approximately higher than 100 W/kg and 0.127, respectively.

Under the condition of the optimal cycle time, when the coolant temperature is controlled at 25 °C, the performance of the system is tested at different regeneration

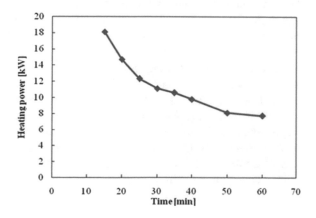

Fig. 5.19 Heating power versus cycle time [29]

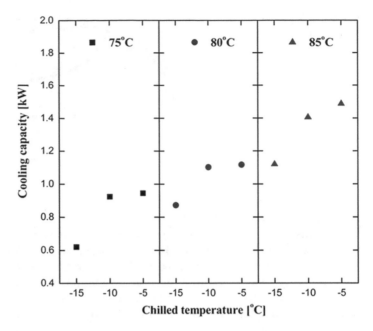

Fig. 5.20 Average cooling capacity versus refrigeration temperature and heat source temperature [29]

temperatures and minimum cooling temperatures, which range from 75 to 85 °C and -15 to -5 °C, respectively.

Figure 5.20 shows that the cooling capacity decreases with the decrease of evaporation temperature and hot water temperature. When the hot water temperature and evaporation temperature are 75 °C and -15 °C respectively, the minimum cooling capacity is 0.62 kW. Figure 5.20 also shows that at different heat source temperatures, the cooling capacity decreases significantly when the evaporation temperature is -15 °C and -10 °C, respectively. It is mainly because of the saturated pressure of the ammonia will decrease with the decreasing refrigeration temperature, which will lead to the deterioration of the mass transfer performance and the sorption performance of the vapor inside the sorbent. The optimal results are obtained from the hot water temperature of 85 °C, for which the cooling capacity increases from 1.12 to 1.49 kW when the evaporating temperature increases from -15 to -5 °C. Figure 5.21 shows the influence of COP and SCP on chilling temperature under three different heat source temperatures. It can be seen that the trend of SCP is similar with the cooling capacity W_{ref}; which increases rapidly as the heating water temperature and evaporating temperature increase. However, the trend of COP is different. COP tends to be more sensitive to evaporating temperature rather than heating water temperature. When the evaporating temperature are -10 and -5 °C, the COP of the system will not change much with the increase of the hot water temperature.

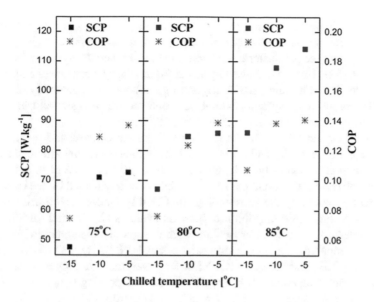

Fig. 5.21 SCP and COP versus refrigeration temperature and heat source temperature [29]

5.1.3   A MnCl₂/CaCl₂-NH₃ Two-Stage Solid Sorption Freezing System for the Refrigerated Truck

Refrigerated trucks are widely employed to transport frozen, fresh and perishable goods. Mechanical vapor compression refrigeration systems are installed in these refrigerated trucks with engine-driven compressors, which will inevitably increase greenhouse gas and particulate emission. For the typical refrigerated transport equipment, it would produce approximately 50 tons of CO_2 annually with a fuel consumption of 0.47 L/h per 1.0 kW refrigerating capacity, operating at an ambient temperature of 30 °C and refrigeration temperature of −20 °C [30]. The sorption refrigeration cycle driven by high temperature exhaust gas could save the energy efficiently if it can be used for refrigerated trucks [31–33].

Sorption air-conditioning systems for the vehicle HVAC (Heating Ventilation Air Conditioning) applications have been investigated. Zhang [34] designed a sorption cooling system driven by the waste heat of a diesel engine, and the sorbent consisted of cylindrical steel double tubes to enhance the heat transfer between the sorbent and the exhaust gas. Jiangzhou et al. [35] and Lu et al. [36] established a sorption air-conditioner for locomotive driver-cabin which utilized zeolite-water as working pair and directly driven by the exhaust gas of internal-combustion engine. Zhong et al. [17] designed a sorption air-conditioning system for heavy-duty truck applications, and the zeolite-water is also applied as its working pair. The system was designed to meet the cooling requirement for idle reduction of long-distance trucks. Abdullah et al. [37] investigated the physical and chemical properties of activated carbon and its feasibility for application in sorption air-conditioning system. Ali et al. [15] proposed a two-stage indirect exhaust heat recovery system for automotive engine, in which an effective lumped parameter model was used to simulate the dynamic behavior of sorption chillers. However, above sorption refrigeration systems are all employed for the vehicle air conditioning. The refrigerating temperature is generally higher than 5 °C. Therefore, these systems cannot be employed for refrigerated trucks whose refrigerating temperature ranges from 0 to −18 °C.

For refrigerated trucks, the temperature of engine exhaust varies with engine load and is between 200 and 500 °C [38]. Simultaneously, the ambient temperature at summer is relatively high, which is above 30 °C. In order to simplify the sorption refrigeration system, the sorption bed is directly heated by the exhaust gas and directly cooled by the surrounding air [39–42]. Under such a critical condition the two-stage solid sorption refrigeration cycle can be applied for adapting to the lower desorption temperature and the higher sorption temperature [43–45]. Jiang et al. [45] compared different working pairs of $CaCl_2/NaBr$-NH_3, $CaCl_2/BaCl_2$-NH_3, $SrCl_2/BaCl_2$-NH_3 and $SrCl_2/NH_4Cl$-NH_3. However, they are only applicable when the heat source temperature is below 100 °C and the cooling temperature is below 30 °C. A novel working pair of $MnCl_2/CaCl_2$-NH_3 is proposed [46], and this working pair is applicable under the high desorption temperature of refrigerated trucks.

In this section, a system is established particularly for the refrigerated truck with the rated power of 80 kW. The hot air heated by the electric heater is utilized to simulate the engine exhaust gas to drive the two-stage solid sorption freezing system. The system performance is investigated under different resorption time, hot air temperatures and refrigerating temperatures [47]. Finally, the corresponding experimental results are analyzed.

5.1.3.1   System Description

For the refrigerated light truck with the rated power of 80 kW, the rated load capacity of this refrigerated truck is 1500 kg. Refrigerated trucks transport chilled or frozen goods with the required refrigerating temperature generally ranges from −18 to 0 °C. The refrigerating capacity required can be evaluated by

$$Q_{cabin} = \varphi K_{cabin} A_{cabin} \left(T_{ambient, \, env} - T_{cabin, \, in} \right) \tag{5.9}$$

where φ is the safety factor, which is usually 2.5. According to the Agricultural Trade Policy (ATP) regulations, the minimum safety factor is 1.75. Together with a company (CIMC Vehicles (Shan dong) CO., LTD) that produces different kinds of refrigerated trucks, the refrigerating capacity is determined. The coefficient of 2.5 covers the types of applications most commonly encountered, and the coefficient authorizes 20 rear door openings for 5 min each over a period of 8 h. And the coefficient of 2.5 also takes into account the deterioration over time of the K_{cabin} value. K_{cabin} represents the total heat transfer coefficient, 0.35 W/(m^2 K). A_{cabin} is

Table 5.5 Required refrigerating capacity under different refrigerating temperatures [47]

Refrigerating temperature/°C	Required refrigerating capacity/W
0	971.2
−5	1133.1
−10	1295
−15	1456.8
−18	1554

the surface area of the cabin. $T_{ambient,env}$ and $T_{cabin,in}$ are the ambient temperature and the refrigerating temperature, respectively. The $T_{ambient,env}$ of 30 °C is chosen, which is the maximum air temperature (average year) in China.

Based on Eq. 5.9, the required refrigerating capacity under different refrigerating temperatures can be calculated. Table 5.5 shows the required refrigerating capacity will be higher when the refrigerating temperature is lower. For the refrigerating temperature of 0 °C, the required refrigerating capacity is only 971.2 W. But for the refrigerating temperature of −18 °C, the required refrigerating capacity is up to 1554 W.

Generally, the waste heat of the engine is discharged directly into the environment, which is approximately equal to 30% of the engine power. The engine exhaust gas temperature is between 200 and 500 °C. When the engine is at low-medium load, the engine exhaust gas temperature is lower than 300 °C [38].

The working process of the two-stage solid sorption freezing system includes the sorption/desorption process and the resorption process, and the system outputs the refrigerating capacity only at the sorption process of $MnCl_2$ bed.

The schematic diagrams of the $MnCl_2/CaCl_2$-NH_3 two-stage solid sorption freezing system are shown in Fig. 5.22a, b. The components in the system include a middle-temperature salt sorption bed (MTS bed), a high-temperature salt sorption bed (HTS bed), an evaporator, a condenser, five ammonia valves (AV), two air valves (CV), two exhaust valves (EV) and other auxiliary equipment. There are two liquid storage tanks in the sorption freezing system to store the refrigerant ammonia. Composite sorbents of $MnCl_2$ and $CaCl_2$ developed by the matrix of ENG-TSA are filled in the HTS bed and MTS bed, respectively. Additionally, four electric fans are employed in the system.

The detailed working principles of the system are as follows:

(1) Sorption process of the $MnCl_2$ bed and desorption process of the $CaCl_2$ bed. During this process, in Fig. 5.22a exhaust valve EV2, air valve CV1, ammonia valve AV1 and AV2 are opened. The HTS bed is cooled by the surrounding air, while the MTS bed is heated by the engine exhaust gas. The ammonia inside the evaporator evaporates and is sorbed by the HTS bed. At the same time, the ammonia in the MTS bed is desorbed and condenses in the condenser. Then the ammonia flows into the liquid storage tank 1. The evaporation process of the ammonia inside the evaporator generates the refrigeration effect, and the

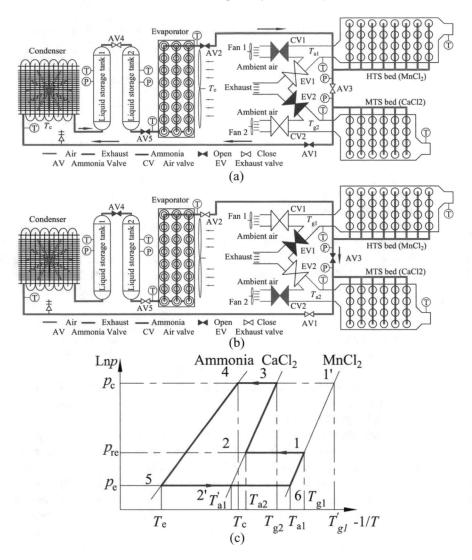

Fig. 5.22 Two-stage solid sorption freezing system: **a** sorption process of MnCl$_2$ bed and desorption process of CaCl$_2$ bed; **b** resorption process between the MnCl$_2$ bed and CaCl$_2$ bed; **c** Lnp-T diagram [47]

chilled air transfers the cooling capacity into the cabin. Compared with the CaCl$_2$-NH$_3$ solid sorption cycle with the evaporating temperature of T_e, the sorption temperature of the two-stage solid sorption cycle rises from the T_{a1}' to T_{a1} (shown in Fig. 5.22c), which has higher adaptability to the high ambient temperature.

(2) Resorption process between the MnCl$_2$ bed and CaCl$_2$ bed. During this process in Fig. 5.22b exhaust valve EV1, air valve CV2, ammonia valve AV3 and AV4

Fig. 5.23 Two-stage solid sorption freezing system on the truck [47]

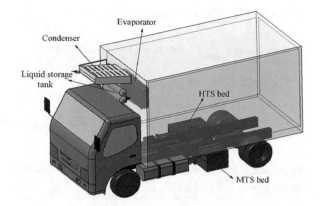

are opened. The HTS bed is heated by the engine exhaust, and the MTS bed is cooled by the surrounding air. The ammonia is desorbed from the HTS bed and flows to the MTS bed. MTS bed sorbs the ammonia and releases the reaction heat to the ambient air. Simultaneously, the liquid ammonia from the liquid storage tank 1 enters the liquid storage tank 2. Compared with the $MnCl_2$-NH_3 solid sorption cycle with the condensing temperature of T_c, the desorption temperature of the two-stage solid sorption cycle decreases from the T_{g1}' to T_{g1} (shown in Fig. 5.22c), which has higher adaptability to the low heat source temperature.

Therefore, the lower desorption temperature of T_{g1} and the higher sorption temperature of T_{a1} make this sorption freezing system suitable for the critical working conditions such as the low-medium load of the truck engine or the high ambient temperature in summer. The installation of the sorption freezing system on the truck is shown in Fig. 5.23. Two sorption beds are installed on both sides of the truck beam. The condenser is installed on the upper side of the nacelle and is cooled by the surrounding air. The evaporator is installed inside the refrigerated cabin, similar to a conventional vapor compression refrigeration system.

5.1.3.2 Performance Calculation and Experimental Results

As mentioned before, the sorption bed is heated by the engine exhaust gas during the desorption process and cooled by the surrounding air during the sorption process. For the desorption process, the heat input is mainly composed of three parts, i.e. the desorption heat of the composite sorbent, the sensible heat of the composite sorbent and the metallic material of the sorption bed.

The instantaneous heat input $Q_{ins, heat}$:

$$Q_{ins, heat} = C_{exh} \cdot m_{exh} \cdot \left(T_{exh, in} - T_{exh, out}\right) \tag{5.10}$$

The average heat input $Q_{ave, heat}$ in a cycle:

$$Q_{ave, heat} = Q_{ave, des} + \frac{1}{t_{cycle}}(C_{ads}M_{ads}\Delta T_{sensible} + M_{met}C_{metl} \cdot \Delta T_{met}) = \frac{\int_0^{t_{cycle}} C_{exh} \cdot m_{exh} \cdot (T_{exh, in} - T_{exh, out})dt}{t_{cycle}} \quad (5.11)$$

The desorption heat $Q_{ave,des}$ is

$$Q_{ave, des} = \frac{1}{t_{cycle}}M_{ads}\Delta x_{ads}\Delta H_{ads} \quad (5.12)$$

where Δx_{ads} and ΔH_{ads} are the cycle sorption quantity and sorption reaction heat, respectively. The Δx_{ads} is:

$$\Delta x_{ads} = \frac{\Delta M_{ref}}{M_{tot,ads}} \quad (5.13)$$

where $M_{tot,ads}$ is total mass of composite sorbent.

The instantaneous refrigerating capacity $Q_{ins, ref}$:

$$Q_{ins, ref} = C_{chilled, air} \cdot m_{chilled, air} \cdot (T_{chilled air,in} - T_{chilled air,out}) \quad (5.14)$$

where the $T_{chilled air, in}$ and $T_{chilled air, out}$ are the chilled air inlet and outlet temperature of the evaporator.

The average refrigerating capacity $Q_{ave sorp, ref}$ in the sorption process of HTS bed:

$$Q_{ave sorp, ref} = \frac{\int_0^{t_{sorp, HTS}} C_{chilled, air} \cdot m_{chilled, air} \cdot (T_{chilled air,in} - T_{chilled air,out})dt}{t_{sorp, HTS}} \quad (5.15)$$

The average refrigerating capacity $Q_{ave, ref}$ in a cycle:

$$Q_{ave, ref} = \frac{\int_0^{t_{cycle}} C_{chilled, air} \cdot m_{chilled, air} \cdot (T_{chilled air,in} - T_{chilled air,out})dt}{t_{cycle}} \quad (5.16)$$

The average refrigerating capacity $Q_{ave sorp, ref}$ is calculated in a half cycle, and the $Q_{ave sorp, ref}$ is almost double of the $Q_{ave, ref}$.

The coefficient of performance:

$$COP = \frac{Q_{ave, ref}}{Q_{ave, heat}} \quad (5.17)$$

where the power consumed by the evaporator fan and the condenser fan isn't considered.

The specific cooling power per kilogram sorbent can be expressed as:

$$SCP = \frac{Q_{\text{ave, ref}}}{M_{\text{tot, ads}}} \tag{5.18}$$

The two-stage solid sorption freezing system is shown in Fig. 5.24a. The HTS bed includes 120 unit tubes and the MTS bed includes 80 unit tubes. Composite sorbents are filled in the unit tube. The mass ratio between the metal chloride and the ENG-TSA is 5:1, and the bulk density of the composite sorbent is about 500 kg/m^3.

During the heating process of the sorption beds, the maximum power of hot air heated by the electric heater is 21 kW to simulate engine exhaust gas. The hot air temperature of the electric heater can be adjusted from the ambient temperature to 280 °C with an accuracy of ±3 °C. Two centrifugal fans are used to cool the two sorption beds separately to simulate the cooling process of the sorption bed

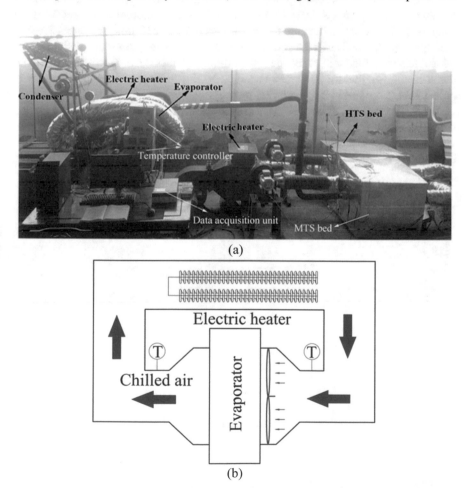

Fig. 5.24 Two-stage solid sorption freezing system: **a** test rig; **b** the method to measure the refrigerating capacity [47]

when a truck is traveling. In order to balance the refrigerating capacity, two electric heaters controlled by temperature controllers are installed in the return-air pipe of the evaporator, which are shown in Fig. 5.24b. Temperature sensors of PT100 and pressure transmitters of YSZK-31 are mounted in the experimental system and their accuracy are ±0.1 °C and 0.3%R respectively.

In order to figure out whether the two-stage solid sorption freezing system could meet required refrigerating capacity of the refrigerated truck at the low-medium load of the engine or not, a series of corresponding experiments are managed.

In the first experiments, the hot air temperature is controlled at approximately 230 °C, and the ambient temperature is approximately 30 °C. The air inlet and outlet temperature of sorption beds at a typical operational condition is shown in Fig. 5.25a, b. For HTS beds, the sorption process between HTS and MTS beds occurs from 0 to 120 min, and from 120 to 240 min HTS sorbs the refrigerant ammonia from the evaporator. However, in the case of the MTS bed, after 210 min, the liquid level in the liquid storage tank 1 of the condenser is almost constant through the sight glass, which means that there is no refrigerant ammonia desorbed from the MTS bed to the condenser. After that the MTS bed is cooled by the ambient air right now, which is beneficial for the following resorption process.

Under the conditions above, the working pressure of two sorption beds is also shown in the Fig. 5.25c. For the HTS bed, only resorption and sorption occur. The working pressure of the HTS bed is at a relative low value ranging from 0.083 to 0.729 MPa. For the MTS bed, the desorption occurs between the sorption bed and

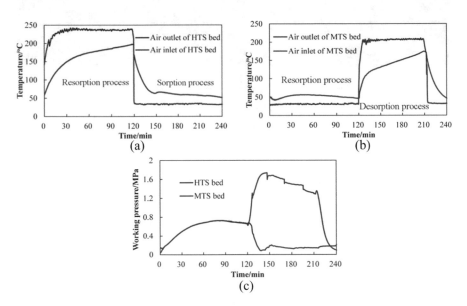

Fig. 5.25 Performance of sorption beds: **a** air inlet and outlet temperature of HTS bed; **b** air inlet and outlet temperature of MTS bed; **c** working pressure [47]

the condenser. The working pressure of the MTS bed is much higher and ranges from 0.140 to 1.734 MPa.

First the performance of the system is investigated under different resorption time of 60, 80, 100 and 120 min with corresponding hot air temperature and refrigerating temperature are 230 °C and −5 °C, respectively. The effective refrigerating time is studied, which refers in particular to the time that the chilled air outlet temperature is the setting temperature such as −5 °C or −10 °C. The volumetric flow rate of hot air and cooling air are 630 and 900 m³/h.

Figure 5.26 shows the instantaneous cooling capacity at −5 °C for different resorption times. Taking the resorption time of 100 min as an example (Fig. 5.26c), when the HTS bed begins to sorb the refrigerant ammonia gas, the outlet temperature of the cold air evaporator gradually decreases from 18 to −5 °C. Then an electric heater in the return duct heats the cold air and the cold air inlet temperature gradually increases. Instantaneous cooling capacity can be calculated, as shown in Fig. 5.26c. Figure 5.26 shows that the effective cooling time increases from 52 min (Fig. 5.26a) to 92 min (Fig. 5.26d) as the resorption time increases, and more ammonia is desorbed from the HTS bed.

Based on the Eq. 5.15, the average refrigerating capacity for the sorption process of HTS bed at different resorption time can be calculated, which is shown in Fig. 5.27a. *COP* and *SCP* are shown in Fig. 5.27b. The average refrigerating capacity (Fig. 5.27a) are almost constant when the resorption time is longer than 80 min, and they are 3.05,

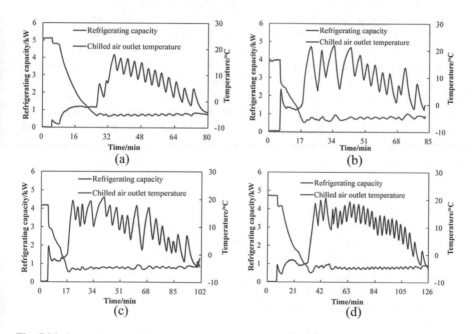

Fig. 5.26 Instantaneous refrigerating capacity with the chilled air outlet temperature of −5 °C under different resorption time of: **a** 60 min; **b** 80 min; **c** 100 min; **d** 120 min [47]

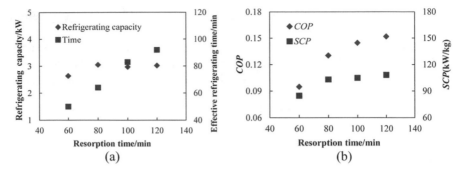

Fig. 5.27 Performance for different resorption time: **a** the average refrigerating capacity in the sorption process of HTS bed and the effective refrigerating time; **b** COP and SCP [47]

2.97 and 3.02 kW under the resorption time of 80, 100 and 120 min. Figure 5.27b indicates that the maximum COP of 0.16 is obtained at the resorption time of 120 min. The SCP at different resorption time of 80, 100 and 120 min are also almost constant, which is about 105 W/kg.

The temperature of the engine exhaust varies with the load and speed of the truck. When the engine is at low-medium load, the engine exhaust gas temperature is below 300 °C. Therefore, different hot air temperatures of 210, 230, 250, 270 °C are used to drive the two-stage solid sorption freezing system. The resorption time and refrigerating temperature are controlled at about 80 min and −5 °C, respectively. The instantaneous refrigerating capacity under different hot air temperature is shown in Fig. 5.28. Based on the Eq. 5.15, the average refrigeration capacity of the HTS bed for the sorption process at different hot air temperature is calculated. With the increase of hot air temperature, the average refrigeration capacity during the sorption process of the HTS bed is almost constant, about 3.0 kW, but the effective refrigeration time increases rapidly from 60 to 98 min, as shown in Fig. 5.29a. The increase of the effective cooling time is related to the increase of the driving temperature potential, which is the difference between the temperature of the heat source and the equilibrium desorption temperature. When the difference is larger, the heat transfer performance is better as well and more ammonia is desorbed from the HTS bed under the same resorption time. The COP and SCP under different hot air temperature are shown in the Fig. 5.29b. Both the highest COP and the highest SCP are obtained at a hot air temperature of 270 °C, which are 0.15 and 108 W/kg, respectively.

Compared the average refrigerating capacity in the sorption process of HTS bed and the effective refrigerating time under the resorption time of 80 min and hot air temperature of 270 °C (Fig. 5.29a) with those under the resorption time of 120 min and hot air temperature of 230 °C (Fig. 5.27a), it shows that they are almost the same. Such a result indicates that a high hot air temperature could effectively shorten the resorption time.

According to the Table 5.5, the required refrigerated capacity for this refrigerated truck is about 1.13 kW under the refrigerating temperature of −5 °C. Based on the Eq. 5.16, the average refrigerating capacity of the two-stage solid sorption freezing

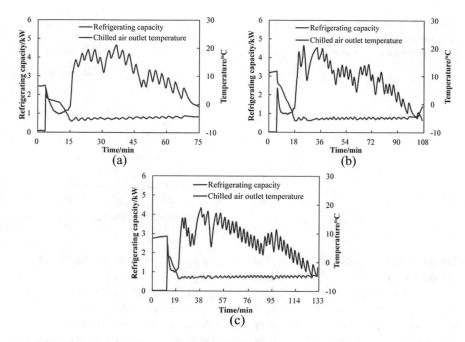

Fig. 5.28 Instantaneous refrigerating capacity under different hot air temperatures of: **a** 210 °C; **b** 250 °C; **c** 270 °C [47]

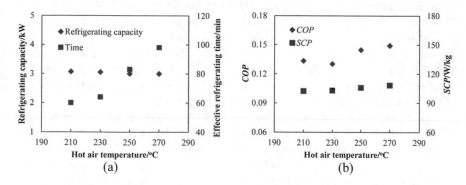

Fig. 5.29 Performance under different hot air temperatures: **a** the average refrigerating capacity in the sorption process of HTS bed and the effective refrigerating time; **b** COP and SCP [47]

system in a cycle versus resorption time is analyzed, which is shown in Fig. 5.30. It shows that most conditions can satisfy the requirement except the average refrigerating capacity of 1.06 kW gotten at the resorption time of 60 min and hot air temperature of 230 °C. For the same sorption time of 80 min, as the temperature of the hot air increases, the average cooling capacity in a cycle increases rapidly. The maximum average cooling capacity is 1.52 kW for the sorption time of 80 min and a

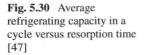

Fig. 5.30 Average
refrigerating capacity in a
cycle versus resorption time
[47]

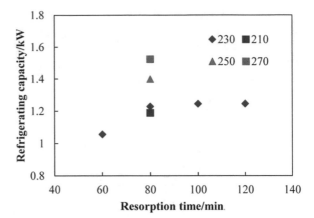

hot air temperature of 270 °C, Therefore, this two-stage solid sorption refrigeration system can meet the requirements for transporting the fresh goods.

For such a cycle there is no refrigerating capacity output during the resorption process, in order to verify if the internal temperature of the cabin will remain almost constant during the entire cycle, the operation condition of 230 °C heating temperature and 80 min resorption time is selected and analyzed. Under such a condition the average refrigerating capacity in a cycle obtained is 1.23 kW, which is higher than the required refrigerating capacity of about 1.13 kW shown in Table 5.5. Simultaneously, it is assumed that the refrigerated truck is utilized to carry the diary product of 1500 kg. As shown in Fig. 5.27a, the average refrigerating capacity in the sorption process of HTS bed is 3.05 kW, and the diary product stores the extra refrigerating capacity in this process. The temperature of the diary product decreases gradually from 2 to 0.8 °C. During the resorption process, the heat capacity of diary product supplies the refrigerating capacity. As a result, the temperature increases from 0.8 to 2 °C, which is acceptable for the storage of dairy products.

Sometimes, transporting frozen food requires a refrigeration temperature between −10 and −18 °C. Therefore, the performance of the two-stage system is tested at different refrigeration temperatures of −10, −15 and −18 °C. Under the chilled air outlet temperature of −5 °C and the hot air temperature of 230 °C, the performances for different sorption time of 60, 80, 100 and 120 min are analyzed. The analysis shows that the maximum *COP* and effective refrigerating time are obtained at the resorption time of 120 min. Therefore, the resorption time of 120 min is selected under the hot air temperature of 230 °C. In addition, the refrigerating temperature of 0 °C is also studied for transporting fresh goods, which is shown in Fig. 5.31a. The results show that the cold air outlet temperature of the evaporator fluctuates greatly in the initial stage. That is because the two electric heaters installed at the return-air pipe of the evaporator cannot balance the refrigerating capacity which is too large due to the high refrigeration temperature. Corresponding instantaneous refrigerating capacity under different refrigerating temperatures is also shown in Fig. 5.31. The lower is the refrigerating temperature, the lower is the refrigeration capacity.

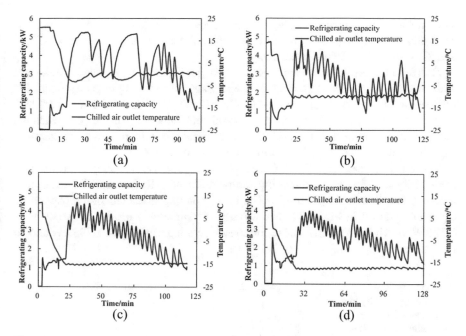

Fig. 5.31 Instantaneous refrigerating capacity under different refrigerating temperatures of: **a** 0 °C; **b** −10 °C; **c** −15 °C; **d** −18 °C [47]

Based on the Eq. 5.15, the average refrigerating capacity in the sorption process of HTS bed is calculated, which is shown in Fig. 5.32a. The average refrigerating capacity decreases gradually from 3.75 to 2.3 kW when the chilled air outlet temperature of the evaporator decreases from 0 to −18 °C. Simultaneously the effective refrigerating time in Fig. 5.32a increases slowly because of the critical mass transfer performance under the lower saturated pressure of the refrigerant ammonia, i.e. the

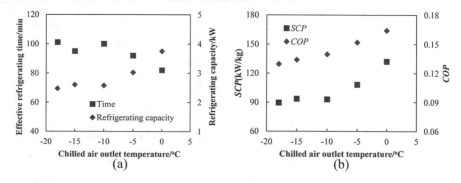

Fig. 5.32 Performance under different refrigerating temperatures: **a** the average refrigerating capacity in the sorption process of HTS bed and the effective refrigerating time; **b** COP and SCP [47]

lower refrigerating temperature. Figure 5.32b shows that both *COP* and *SCP* increase gradually with the increase of the refrigerating temperature, and they are 0.13–0.16 and 89.6 W/kg–132 W/kg, respectively.

Based on the Eq. 5.16, the average refrigerating capacity in a cycle is calculated and shown in Fig. 5.33. It shows that the average refrigerating capacity under the refrigerating temperature of 0 °C, −10 °C, −15 °C, 18 °C are 1.42 kW, 1.09 kW, 1.06 kW and 1.04 kW, respectively. Compared with Table 5.5, the data can reach the required refrigerating capacity of 0 °C, but it cannot reach the required refrigerating capacity of the refrigerating temperature lower than −10 °C.

The hot air temperature of 270 °C is employed to drive the two-stage solid sorption freezing system for the refrigerating temperature of −10 °C, and the resorption time is controlled at about 80 min. The corresponding instantaneous refrigerating capacity is shown in Fig. 5.34. The average refrigerating capacity in the cycle is 1.32 kW, which is higher than the required refrigerating capacity of 1.295 kW shown in Table 5.5.

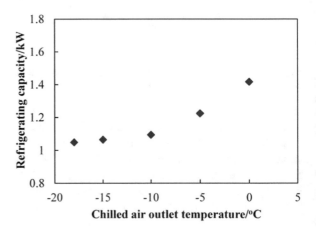

Fig. 5.33 Average refrigerating capacity in a cycle versus the chilled air outlet temperature [47]

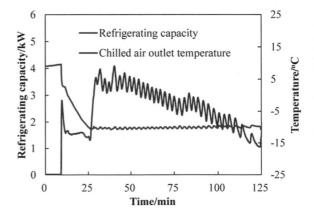

Fig. 5.34 Instantaneous refrigerating capacity with refrigerating temperature of −10 °C and hot air temperature of 270 °C [47]

Table 5.6 Refrigerating capacity obtained under different operation conditions [47]

Refrigerating temperature/°C	Required RC/kW	Hot air temperature/°C						
		210	230				250	270
		Resorption time/min						
		80	60	80	100	120	80	80
		RC/kW	RC/kW	RC/kW	RC/kW	RC/kW	RC/kW	RC/kW
0	0.97					1.42		
−5	1.13	1.19	1.06	1.23	1.24	1.25	1.4	1.52
−10	1.29					1.09		1.32
−15	1.45					1.06		
−18	1.55					1.05		

However, the analysis shows that the average refrigerating capacity in a cycle can reach the required refrigerating capacity only when the hot air temperature is heated to 330 °C for the refrigerating temperatures of −15 and −18 °C. It is because the resorption time is effectively shortened by high heat source temperature. Consequently, this two-stage solid sorption freezing system cannot reach the required refrigerating capacity with the lower refrigerating temperature when the engine is at low load. When the engine is at medium–high load, the engine exhaust gas temperature is easily to be higher than 300 °C, and the heat of the exhaust gas is relatively large at the same time. If the refrigerated truck is utilized for the long-distance transportation of frozen goods, the engine is always at relatively heavy load. At this time the engine exhaust gas temperature is high enough to make this two-stage solid sorption freezing system reach the required refrigerating capacity even under the refrigerating temperature of −18 °C.

In order to clearly demonstrate the refrigerating capacity obtained under different operation conditions, the Table 5.6 has been added.

5.1.3.3 Optimization Design of the Freezing Prototype System

Consider the employment of the $MnCl_2/CaCl_2$-NH_3 two-stage solid sorption freezing cycle, experimental results show that the refrigerating capacity obtained could satisfy requirement for transporting fresh goods even under the low engine load condition in summer, but the system has the following problems:

(1) The system can reach the required refrigerating capacity only for transporting fresh goods at the refrigerating temperature of −5 to 0 °C. However, it is difficult to satisfy the required refrigerating capacity when the refrigerating temperature is lower than −10 °C.
(2) Since there is no refrigerating capacity output in the resorption process, and the resorption time (120 min) accounts for about half of the cycle time of the

refrigeration system (240 min). Therefore, selecting the optimal resorption time can improve the cold storage effect of the goods.

(3) The total mass of the system is about 260 kg, and the performance will be more improved if the mass of the system can be decreased.

In order to properly solve the above problems, a series of measures are conducted to optimize the design of the system [48]. Simultaneously, corresponding experimental research on the system has been carried out.

The structure of the sorption beds composed of many unit tubes are similar to the two-stage solid sorption refrigeration system previously constructed by Gao et al. [47]. In the previously constructed refrigeration system, the average cooling capacity of the system is relatively low because the long cycle time is up to 240 min. The insufficient heat exchange between the unit tube and air in the sorption bed is the main reason for the long cycle time. Furthermore, the ideal air flow path in the previously constructed HTS bed is analyzed, as shown in Fig. 5.35a. It can be seen from Fig. 5.35a that the effective heat transfer area of each unit tube surface is quite restricted. The previous aligned arrangement of the unit tube is changed to staggered arrangement, which can significantly enhance the heat transfer effect between the unit tube and air. At the same time, the optimized air path in the HTS bed is analyzed as shown in Fig. 5.35b. From Fig. 5.35b the contact area between each unit tube and air is increased due to the fact that each unit tube is almost surrounded by air, which strengthens the heat transfer effect.

The composite sorbent is pressed into the unit tube by the mass ratio of ENG-TSA to metal chloride of 1:5, and its packing density is about 500 kg/m^3.

As described, the sorption bed is cooled by ambient air in the sorption process and heated by engine exhaust gas in the desorption process. The thermal energy input used to heat each unit tube in the desorption process is mainly composed of desorption heat of the composite sorbent, the metallic materials of the unit tube and the sensible heat of the composite sorbent. The average thermal energy input of each unit tube in the desorption process $Q_{\text{ave, tube, heat}}$ is:

$$Q_{\text{ave, tube, heat}} = KA_{\text{tube}} \Delta T = \frac{1}{t_{\text{des}}} \left(M_{\text{tube, sor}} \Delta x_{\text{sor}} \Delta H_{\text{sor}} + C_{\text{sor}} M_{\text{tube, sor}} \Delta T_{\text{sensible}} + M_{\text{tube, met}} C_{\text{met}} \cdot \Delta T_{\text{met}} \right)$$

(5.19)

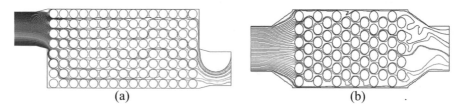

(a)   (b)

Fig. 5.35 Flow path lines of the air in the HTS bed: **a** the previously constructed HTS bed; **b** the later optimized HTS bed [48]

where $M_{tube,sor}$ is the mass of the composite sorbent filled in the unit tube, and $M_{tube,met}$ is the metallic mass of the unit tube. Δx_{sor} and ΔH_{sor} are the cycle sorption quantity and reaction heat, respectively. Δx_{sor} is:

$$\Delta x_{sor} = \frac{\Delta M_{tube, ref}}{M_{tube,sor}} \tag{5.20}$$

where $\Delta M_{tube,ref}$ is the refrigerant mass that can be sorbed by the metal chloride/ENG-TSA composite sorbent in the unit tube.

When the automobile exhaust gas scours the outer wall of the unit tube, the total thermal resistance consists of the thermal resistance of the composite sorbent itself, the thermal resistance of the metal tube, the fouling resistance of the tube wall, and the convective heat transfer resistance between the exhaust gas and the outer surface of the unit tube. The Equation is:

$$\frac{1}{K} = \frac{1}{h_{exh}} + R + \frac{A_{tube} \ln(r_{tube}/r_{sor,ext})}{2\pi \lambda_{tube} l} + \frac{A_{tube} \ln[r_{sor}/(r_{sor,ext} - r_{sor,inn})]}{2\pi \lambda_{sor} l} \tag{5.21}$$

where the λ_{sor} is the thermal conductivity of the metal chloride/ENG-TSA composite sorbent, which is 18.7 W/(m K) with the packing density of 500 kg/m^3 [49].

The convection heat transfer coefficient of the automobile exhaust gas is:

$$h_{exh} = \varepsilon \times 0.35 \times \left(\frac{S_1}{S_2}\right)^{0.2} Re_{exh}^{0.6} Pr_{exh}^{0.36} \left(Pr_{exh}/Pr_{tube}\right)^{0.25} \times \lambda_{exh}/(2 \times r_{tube}), \frac{S_1}{S_2} \leq 2 \tag{5.22}$$

The average refrigerating capacity attain from a unit tube:

$$Q_{ave, tube, ref} = \frac{\Delta M_{tube,ref} \times L}{t_{cyc}} \tag{5.23}$$

Furthermore, the thermodynamic and heat transfer of the unit tube is calculated, and the design of the unit tube and sorption bed is optimized. The results show that the number of unit tubes in the MTS bed and HTS bed decrease from 80 to 55 and from 120 to 66, respectively. Consequently, the total mass of the whole sorption beds system is significantly reduced.

Moreover, the pressure drop of the sorption bed increases suddenly because the unit tubes in the sorption bed are the staggered arrangement instead of the aligned arrangement. Therefore, the pressure drop of the sorption bed can be effectively optimized by adjusting the spacing S_1 between two unit tubes. Figure 5.36a is the test bench for accurately measuring the pressure drop between the inlet and outlet of the sorption bed. Figure 5.36b shows the layout of the wooden club unit pipe, and the voltage controller regulates the air volume of the fan.

Fig. 5.36 Test rig for measuring pressure drop between the air inlet and outlet of the sorption bed:
a experimental system; **b** arrangement mode of unit tubes [48]

Different spacing of 57 and 52 mm is chosen, and test results are shown in Fig. 5.37.
Both the pressure drop and the volumetric flow increase with the rising voltage. Under
the same voltage, the pressure drop at the spacing of 57 mm is lower than that at the
spacing of 52 mm, but the volumetric flow at the spacing of 57 mm is higher than
that at the spacing of 52 mm. When the voltage is 225 V, the pressure drop at the
spacing of 57 and 52 mm is 363 and 559 Pa, and the corresponding volumetric flow
is 1765 and 1514 m^3/h. Considering that no fan is installed at the sorption bed, the
small pressure drop is beneficial for the system. As a result, the spacing of 57 mm is
chosen. Consequently, the size of the MTS bed and HTS bed is 699 × 520 × 341 mm
and 799 × 520 × 341 mm. The optimized HTS bed is shown in Fig. 5.38.

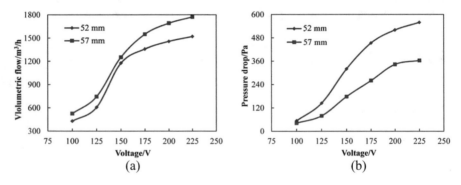

Fig. 5.37 Test results versus voltage regulating the air volume of the fan: **a** pressure drop between
the air inlet and outlet of the sorption bed; **b** volumetric flow [48]

Fig. 5.38 HTS bed (MnCl$_2$
bed [48])

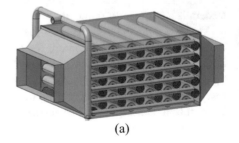

Two sorption beds in the sorption freezing system on the truck are mounted on two sides of the truck beam. The installation of evaporator, the liquid storage tank and the condenser is similar with that shown in Fig. 5.23.

5.1.3.4 Experimental Results of the Optimized System

The $MnCl_2/CaCl_2$-NH_3 two-stage solid sorption freezing experimental system after optimization is established as shown in Fig. 5.39.

To determine the performance of the optimized system, different hot air temperatures of 210 and 230 °C, different resorption time of 20, 30 and 40 min, and different refrigerating temperatures of 0, −5, −10, −15, −20 and −25 °C are studied. As the same time, the performance of the optimized system is compared with that of the previously constructed system.

The temperature changes of air inlet and outlet of sorption bed under typical operation conditions are shown in Fig. 5.40a. The ambient temperature and hot air temperature are 15 and 210 °C. For the HTS bed, the desorption process takes place from 0 to 30 min between the HTS bed and the MTS bed, and the sorption process occurs in the HTS bed between 30 to 65 min.

The working pressure of the sorption beds is shown in Fig. 5.40b. For the HTS bed, the sorption process occurs between $MnCl_2$ bed and $CaCl_2$ bed. Therefore, the working pressure is at a relatively low value, ranging from 0.083 to 0.479 MPa. According to the equilibrium line of the reaction between ammonia and chloride

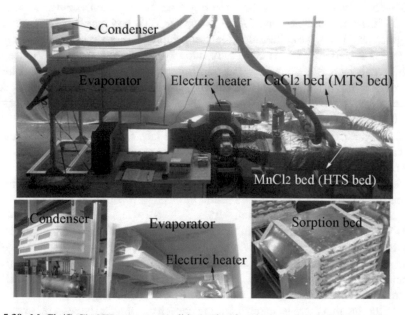

Fig. 5.39 $MnCl_2/CaCl_2$-NH_3 two-stage solid sorption freezing experimental system [48]

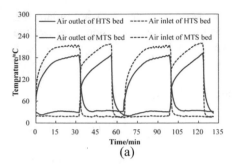 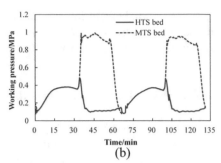

Fig. 5.40 Performance of sorption beds: **a** air inlet and outlet temperature of the HTS and MTS bed; **b** working pressure [48]

[27], the relatively low equilibrium temperature corresponds to the relatively low working pressure; i.e., such a process effectively reduces the required desorption temperature. At the same time, for the MTS bed, the desorption process happens between the sorption bed and the condenser. As a result, the working pressure ranges from 0.110 to 0.973 MPa, which is relatively higher.

As mentioned above, the refrigeration temperature generally ranges from -18 to $0\ °C$. Different refrigeration temperatures of freezing system are studied experimentally. The hot air temperature is set at about $210\ °C$ and the cyclic half period is approximately 30 min.

The real time refrigerating capacity for different freezing temperatures of $0, -5, -10, -15, -20$ and $-25\ °C$ is shown in Fig. 5.41. Setting the freezing temperature of $-10\ °C$ as an example (Fig. 5.41c), when the HTS bed starts to sorb ammonia refrigerant at 35 min, the chilled air outlet temperature of the evaporator drops rapidly from -2.1 to $-10\ °C$, which takes only one min. After that, the inlet temperature of chilled air increases gradually. The real time refrigerating capacity is calculated according to Eq. 5.14, and the calculation results are shown in Fig. 5.41c. A large amount of refrigerating capacity output only occurs during the sorption process of HTS bed.

Based on Eq. 5.16, the average refrigerating capacity of the whole cycle is calculated according to different cooling temperature, as shown in Fig. 5.42a. Because of the critical mass transfer performance under the lower evaporating pressure of the ammonia refrigerant, the average refrigerating capacity decreases gradually with the decreasing cooling temperature. As a result, the average refrigerating capacity at $0\ °C$, $-5\ °C, -10\ °C, -15\ °C, -20\ °C$ and $-25\ °C$ is 2.46 kW, 2.32 kW, 2.06 kW, 1.94 kW, 1.69 kW and 1.63 kW, correspondingly. Obviously, the experimental results are larger than the refrigerating capacity shown in Table 5.5, which can meet the design requirements. In addition, COP and SCP data at different refrigeration temperatures are also obtained, as shown in Fig. 5.42b. When the refrigeration temperature is $0\ °C$, the maximum COP is 0.288 and the maximum SCP is 186.4 W/kg.

Considering that there is no refrigerating capacity output in the resorption process between the HTS bed and the MTS bed, the resorption time accounts for about

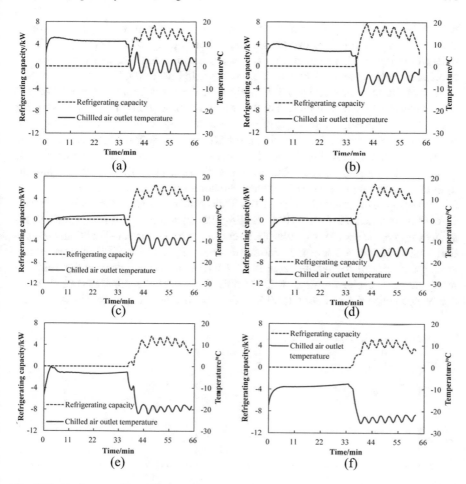

Fig. 5.41 Real time refrigerating capacity under different cooling temperatures of: **a** 0 °C; **b** −5 °C; **c** −10 °C; **d** −15 °C; **e** −20 °C; **f** −25 °C [48]

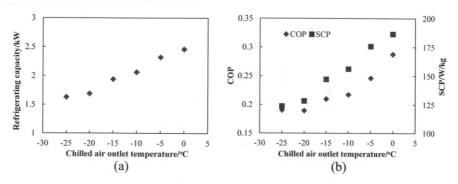

Fig. 5.42 Performance under different refrigerating temperatures: **a** average refrigerating capacity; **b** COP and SCP [48]

half of the cycle time of the refrigeration system. An appropriate resorption time is extremely important to improve the storage effect of the goods. Therefore, different resorption time of 40, 30 and 20 min are experimentally studied, which corresponds to different cycle time of 80, 65 and 45 min shown in Fig. 5.43a. Experimental results show that the refrigerating capacity obtained at the resorption time of 40 min and 20 min can meet the requirement. For the resorption time of 20 min, the refrigerating capacity at the cooling temperature of -15, -10 and -5 °C is 1.87, 1.9 and 2.15 kW. In addition, the refrigeration capacity, the COP and the SCP of different resorption time are compared. From Fig. 5.43b–d, it can be seen that the maximum refrigerating capacity, COP and SCP can be obtained when the resorption time is 30 min, and the optimum resorption time is 30 min when the hot air temperature is 210 °C.

According to the regulations of Carrier Transicold distributor, the cooling temperature of frozen meat foods is -10 °C. Setting the transportation of frozen meat foods as an example, the operating conditions of cooling temperature of -15 °C and resorption time of 30 min are selected. Under this operating condition, the average refrigerating capacity of the whole cycle is 1.94 kW, which is larger than the 1.45 kW refrigerating capacity required in Table 5.6. The selected refrigerated truck has a deadweight capacity of about 1500 kg. The specific heat capacity of frozen meat

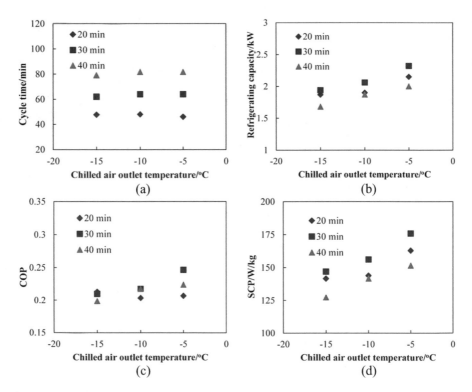

Fig. 5.43 Refrigeration performance under different resorption time: **a** cycle time; **b** average refrigerating capacity; **c** COP; **d** SCP [48]

foods is approximately 3.2 kJ/(kg K). The average refrigerating capacity of the HTS bed is about 3.84 kW, as shown in Fig. 5.41d. Therefore, when the resorption time is approximately 31 min, the frozen meat foods store additional refrigerating capacity in the process, and the temperature change of frozen meat foods is

$$DT = (3.84 \text{ kW} - 1.46 \text{ kW}) \times (31 \times 60 \text{ s})/(1500 \text{ kg} \times 3.2 \text{ kJ}/(\text{kg K})) = 0.9 \text{ K}.$$

Thus, the temperature fluctuation of frozen meat is 0.9 °C, such a small temperature difference range during storage is acceptable.

The hot air temperature of driving two-stage solid sorption refrigeration system is set at 230 °C and the half cycle period is 30 min. Figure 5.44 compares the experimental results with those obtained at the hot air temperature of 210 °C. It can be seen from Fig. 5.44a–c that the refrigerating capacity, COP and SCP of the two-stage solid sorption refrigeration system are nearly the same. However, the effective cooling time shown in Fig. 5.44d has increased from 26 to 30 min. The effective cooling time is defined as the time when the outlet temperature of chilled air reaches the set temperature, such as −10 °C or −15 °C.

Table 5.7 shows the refrigerating capacity (RC) of two-stage solid sorption refrigeration system under different operating conditions.

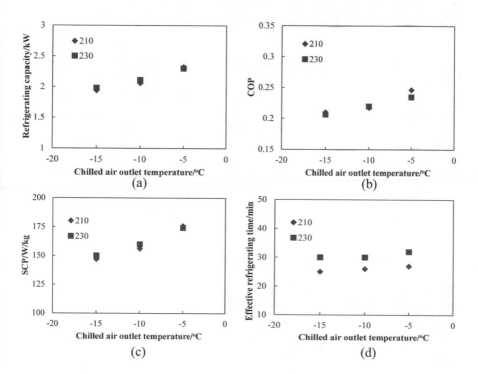

Fig. 5.44 Performance under different hot air temperature: **a** average refrigerating capacity; **b** COP; **c** SCP; **d** effective refrigerating time [48]

Table 5.7 Refrigerating capacity (RC) obtained under different operating conditions [48]

Refrigerating temperature/°C	Required RC/kW	RC/kW			
		Resorption time (min)/Hot air temperature (°C)			
		20/210	30/210	40/210	30/230
0	0.97		2.46		
−5	1.13	2.15	2.32	2	2.28
−10	1.29	1.9	2.06	1.87	2.11
−15	1.45	1.87	1.94	1.68	1.98
−20	1.62		1.69		
−25	1.78		1.63		

5.1.3.5   Performance Comparison of Solid Sorption and Absorption Refrigerating Systems

Solid sorption refrigerating system is compared with the ammonia-water sorption refrigeration system. Table 5.8 shows the experimental performance of ammonia sorption chillers. The ammonia-water diffusion sorption chillers are often called small capacity chillers, which are used pressure compensated auxiliary gas (helium or hydrogen). Because there is no moving part in diffusion sorption chillers, they can

Table 5.8 Experimental performance of some ammonia-water absorption chillers [48]

Author	Mixture	Heat source/Temperature (°C)	Evaporation temperature/°C	COP	Refrigerating capacity/kW
Bazzo et al. (2013)	Ammonia-water	Steam/160.9	−9.5	0.35	7
Said et al. (2016)	Ammonia-water	Hot water/129	−7	0.3	4.8
ACF60 LB (2016)	Ammonia-water	Natural gas/>500	−10	0.48	13.3
Jakob et al. (2008)	Ammonia-water-helium	Electricity/129	6	0.12	0.7
Mazouz et al. (2014)	Ammonia-water-hydrogen	Electricity/185	−14	0.12	0.005
Rêgo et al. (2014)	Ammonia-water-hydrogen	Engine exhaust gas/200	−7	–	–

Note "–" indicates that the data isn't shown in the paper

operate with the absolute noiselessness, which have been commercially developed for the hospital rooms and the hotel mini-bars. But both the COP and the refrigerating capacity are comparably low. The COP is approximately 0.12 and the refrigerating capacity is usually less than 1 kW. Hence, the diffusion sorption chillers are hard to meet the demand of refrigerated trucks. A single-stage ammonia-water sorption chiller also known as large capacity chillers, for which both the COP and the refrigerating capacity are above 0.3 and 5 kW, respectively. The technology of the spray falling film evaporation is used in the refrigerating system. When trucks vibrate continually at driving conditions, the performance will be deteriorated by vibration, because the liquid that is sprayed cannot exchange heat with the tube very well. At the same time, the solution pump makes the system more tedious and also consumes some electricity. In addition, in consideration of the cost and flow of solution pump, it is difficult to find a reasonable solution pump for refrigerated trucks in the market. Consequently, the large capacity ammonia-water chiller is also not appropriate for refrigerated trucks at present.

5.2 Solid Sorption Cycle for Water Production

Freshwater, which is highly dependent on the biology of the earth, accounts for only 3% of the total water resources, which directly affects the development of industry and economy, and also determines the distribution of life [50]. In remote areas such as desert islands lacking corresponding power infrastructure [51], freshwater cannot be produced in desalination plants [52]. In addition, due to less precipitation and large evaporation in the arid desert area, the surface freshwater resources in the area are extremely scarce [53].

As we all know, water vapor is entrained by the airflow after rising from the sea to form freshwater. It accounts for 0.04% of the total freshwater in the world, and its saturation temperature is between 10 and 40 °C. 9.6–55.3 g of freshwater can be extracted per cubic meter of dry air [54]. Therefore, air-to-water technology is a flexible and prospective way to meet the needs of people in remote island areas with high moisture content of air [55]. There are three ways to extract water from the air, among which heat pump is used for surface cooling [56] or radiation cooling [57] with the temperature differences to reach the dew point, which is often accompanied by high energy consumption. Because most chemical reagents are not safe enough, water vapor concentrators using liquid solutions [55] are usually toxic. The solid sorption technology has the advantages of safety, energy conservation and reliability, which has broad application prospects and market potential [58].

The technology, which is employed to sorb water vapor in cooling phase and then the heating phase is utilized to desorb water extract to water, is called sorption air-to-water technology. However, the circulating water sorption performance of sorbent seriously affects the water intake performance of air–water sorption device. This means that the sorbent must have high water sorption capacity under natural climate conditions. At the same time, due to the relatively low temperature of solar

heat accumulation in the solar air–water sorption device, it is necessary to optimize the performance of the sorbent and reduce its desorption temperature as far
as possible to obtain a large amount of freshwater. Therefore, for the utilization of
solar heat with lower temperature, reducing the desorption temperature can significantly improve the water intake performance of the device. Through the synthesis of
a large number of solid composite sorbents and in-depth research, it is proved that
this kind of sorbent has outstanding stability, practicability and economy [59]. Silica
gel [60], ordered mesoporous silicate [61] and activated carbon [62] are commonly
used as porous sorption substrates. Solid composite sorbents, such as lithium chloride
[63, 64] and calcium chloride [65, 66], have been widely studied because of their high
cyclic sorption capacity. In this regard, Aristov et al. prepared composite sorbents by
impregnating hygroscopic salts (LiBr [67], $CaCl_2$ [65, 66]) in porous media (SWS-
1L [54], MCM-41 [68], FAM-Z02 [69]) to prepare composite sorbents. However,
most of the materials cannot have both high water sorption and high desorption
performance under the driving of low temperature heat source. In fact, the granular
solid sorbents are separated from each other after the working fluid is desorbed,
which leads to the poor heat transfer performance in the desorption process. In addition, it is very important to optimize the structure of sorption bed. The sorption and
desorption time can be shortened by designing a sorption bed with good heat and
mass transfer performance and high efficiency. Gordeeva et al. used their composite
sorbent materials [67] on two sorption beds, the sorbent mass is 250–350 g, and
the typical sorption equilibrium time is 50–60 h and the desorption time is 30 h
[70]. Later, Ji et al. manufactured a small-scale solar driven freshwater production
system, which used MCM-41 porous matrix and $CaCl_2$ composite sorbent. The cost
of popularization in large-scale units is too high, and the filled sorbent is too small
to obtain a large amount of freshwater [71]. Obviously, the method of directly filling
a large amount of sorbent to prepare the sorption bed cannot significantly shorten
the cycle time due to the poor heat transfer performance among silica gel particles.
Hamed et al. studied the relationship between the mass of water absorbed from the
atmosphere and the sorption area and mass transfer coefficient. However, the mass
transfer channel inside the corrugated bed uses deformable cloth to transport salt
solution, which has low packaging density and low calculation accuracy. Moreover,
the device only produces 1.5 kg freshwater per square meter per day [72]. Meanwhile, Kabeel et al. explored the glass pyramid shape with multi-shelf solar system
to extract water from humid air, and the multi-shelf structure increased the amount
of the sorbents but this method is the stacking method essentially with bad mass and
thermal transform performance [73].

A new type of composite sorbent has been invented [74], and it is consolidating activated carbon felt (ACF [75, 76]) with LiCl [59]. Meanwhile, a sorption
device based on ACF LiCl composite sorbent is manufactured and designed, and the
performance of the system is comprehensively studied.

Research presents that such consolidated composite sorbent has the features of
optimal water uptake performance, super structure stability and excellent desorption
property. The shortest equilibrium desorption time is only 150 min, and the largest
capacity of desorption is 0.65 g/g at 70 °C and 15% relative humidity. At the same

time, consolidating ACF matrix has good structural properties and hard deformation after drying and easy shaping before drying. The heat transfer and mass transfer channels of sorbent bed can be built directly for the excellent structure formability. Meanwhile, the conventional metal mesh channels which increase the extra weight and thermal capacity are no longer needed. At the same time, the wave and flat solidified composited sorbents layered in staggered arrangement are invented as a novel filling mode to constitute the sorption bed structure, since the direct stacking method increases the resistance of gas flow greatly, and also weakens the ability of heat and mass transfer. The ACF-LiCl curing sorbent and its filling method is studied because it can be compactly and efficiently filled on the sorbent bed. Such type of the solid composite sorbent also can expand the direct contact area between sorbent and air, and reduce the flow resistance greatly because of the uniform distribution of the mass transfer channels formed by a large amount of sorbent in the bed.

5.2.1 Principle of the Cycle and System Design

The principle of the semi-open solar-driven sorption air-to-water cycle is shown in Fig. 5.45. There are two working phases:

(1) Sorption phase (Fig. 5.45a): moist air of environmental species flows into the sorption bed at night. The sorbent sorbs the water vapor in the air in the sorption bed, which release the sorption heat into the ambient air. After sorption, dry air flows out of the system.

(2) Desorption and condensation phase (Fig. 5.45b): the recovered air in the unit is heated by solar energy, and then the sorbate is desorbed from the sorbent bed. The water vapor is mixed with air and flows into the condenser, which is cooled by environmental fins. When the temperature of the wet air drops to the dew point, the water is condensed and collected.

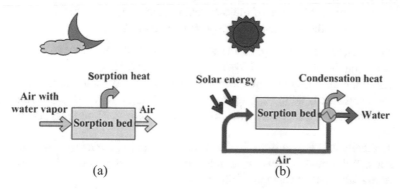

Fig. 5.45 Principle of the solar-driven sorption air-to-water cycle. **a** Sorption phase; **b** desorption and condensation phase [77]

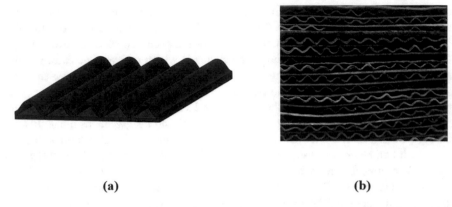

(a)   **(b)**

Fig. 5.46 Structure of the sorbent. **a** Sorbent unit; **b** the layered sorbent units in sorption bed [77]

The new material is solidified ACF matrix with lithium chloride (ASLI) as composite material. The solidified ACF matrix with nano silica particles in the gap of pure ACF fiber has the advantages of firm support frame, no deformation, good heat transfer performance and large specific surface area. As shown in Fig. 5.46a, the unit structure is corrugated and easy to form mass transfer channels. In the sorption bed, the plate-shaped sorbent and corrugated sorbent units are vertically placed layer by layer, as shown in Fig. 5.46b.

The sorption bed can pack 40.8 kg of consolidating sorbents in 0.4 × 0.4 × 0.6 m size as shown in Fig. 5.46, and this structure of design has three main advantages:

(1) The high quality of sorbent filled in the bed ensures that the bed has enough circulating sorption capacity, which is the key to improve the water production capacity.
(2) Due to the structure of carbon fiber, the heat in the air can be efficiently and evenly transferred to the sorbent. In previous studies, the sorbent can be desorbed at the temperature as low as 70–80 °C.
(3) The uniform distribution of mass transfer channels reduces the flow resistance and improves the sorption and desorption performance.

The numerical calculation model is shown in Fig. 5.47. Air flows in the direction of the arrow on the Y-axis (Fig. 5.47a), and the fluid is set to be incompressible. The viscosity of the fluid is considered and the effect of gravity is ignored. ICEM mesh generation software is adopted to mesh the entire fluid domain using a combination of hexahedral and Y-shaped meshes, and to encrypt at the boundary as shown in Fig. 5.47b.

The velocity and temperature in the Y direction are taken as the inlet boundary conditions, and periodic boundary conditions are set in the X-axis direction of the element to realize the expansion of the model.

Reynolds number Equation:

$$\mathrm{Re} = \rho v l / \mu \tag{5.24}$$

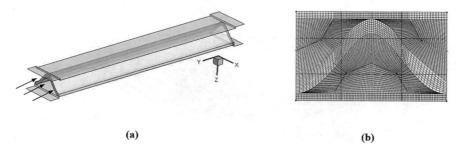

(a)   (b)

Fig. 5.47 CFX simulation of sorbent bed unit, **a** numerical calculation model; **b** grid layout of low velocity turbulence model [77]

where ρ is the density of the fluid, v is the velocity of the fluid, μ is the dynamic viscosity, and l is the characteristic diameter of the irregular cross-section in this model, so l is converted to the hydraulic diameter by:

$$l = 4S/P \qquad\qquad (5.25)$$

where S is the cross-sectional area and P is the wet perimeter length.

The inlet wind speed is set at 2 m/s, and the fluid temperature is 25 °C, which is consistent with the actual working conditions. Therefore, the calculated Reynolds number is 2564, which belongs to low Reynolds number. Therefore, SSTK-ω model is used as the low Reynolds number turbulence model to calculate and analyze the sorption bed unit.

The pressure analysis results are shown in Fig. 5.48a. The pressure drops slightly from the inlet to the outlet, indicating that the flow resistance is very small. At the same time, the differential pressure transmitter is employed to measure the two ends of the single-layer sorbent. The difference between the inlet and outlet is 1 mm water column, i.e. 9.8 Pa, and the simulation value is 8.8 Pa, which is consistent with the experimental results, indicating that the simulation results are reliable. It should be noted that the pressure on the inlet sorbent surface is 9 Pa, which prevents the fluid and forces it to diverge. The pressure on the bell shaped surface changed from 5.8 to 0 Pa, resulting in negative pressure at the two corners of the outlet and sorbent.

The velocity distribution is shown in Fig. 5.48b. The inlet velocity is 2 m/s measured by a hot wire anemometer. Since the velocity in the X and Z directions is negligible, it is the velocity in the Y direction. After entering the unit, the air splits and the speed increases from 2 to 3 m/s, which is due to the decrease of inlet cross-section area.

When the fluid enters the channel, the axial velocity of the center is the fastest, about 4 m/s; the closer to the sorbent wall is, the lower the velocity is, which is approximately 2.5 m/s. At the boundary of sorbent and outlet, the lower angular velocity decreases to 0.5 m/s and the upper angular velocity drops to 1.5 m/s due to the generation of fluid vortices.

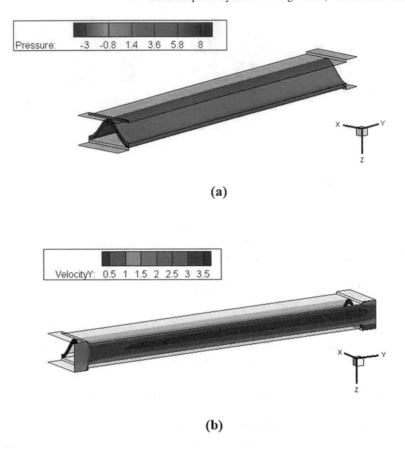

Fig. 5.48 CFX simulation results of the unit structure of sorbent bed using SST k-ω model, **a** pressure; **b** velocity [77]

5.2.2   The Lab System Driven by Electricity

5.2.2.1   The System Design of the Device

As shown in Fig. 5.49a, the four main components of the air-to-water system are heater, sorption bed, condenser and fan. The heating process of solar collector is simulated by using electric heater, and the feasibility of the system is verified. The manufacturing of sorbent is shown in Fig. 5.46b, which can promote the air circulation and make the sorbent quickly extract water and sorb water. Water cooled condenser is adopted. In air circulation, sorption and desorption are used to provide power. Figure 5.49b shows a photo of an air-water device. Automatic temperature controller is used to adjust the temperature output of the heater. Four relative humidity sensors and six PT100 resistors are set at the outlet of sorption bed and condenser respectively.

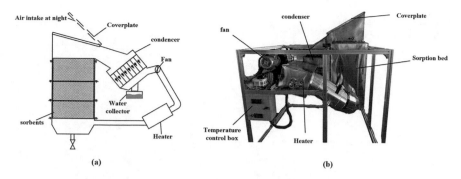

Fig. 5.49 Schematic of electrical heating device based on open-type in sorption and closed-type in desorption cycle, **a** schematic diagram of the device; **b** device diagram [77]

In fact, as shown in Fig. 5.50, the system is open during sorption and closed during desorption. The working stages are as follows:

(1) Sorption stage: the cover plate (Fig. 5.50a) is opened. When the fan is turned on, the air flows into the device and passes through the sorbent from the bottom of the bed. The water vapor in the air is sorbed by the sorbent. The air then flows through the cover plate to the environment.

(2) Desorption stage: the cover plate is closed (Fig. 5.50b), and the air is heated by the electric heater and then flows to the sorption bed. The bed is heated by hot air, so the water vapor is desorbed. After that, the desorbed steam carried by the wet air flows to the condenser, where it condenses into liquid water and is collected by gravity at the bottom of the condenser. The condensed air is drawn out by the fan and passes through the sorption bed again.

The humidity map of desorption phase is shown in Fig. 5.51, and there are four main points in the cycle. Point 1 is ambient air, and the absolute moisture content of 1–2 remains unchanged during heating process. Then the high temperature air enters bed to desorb the sorbents, which is 2–3, where the humidity rises up and

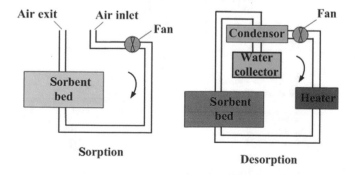

Fig. 5.50 Working diagram of sorption and desorption cycle [77]

Fig. 5.51 Psychrometric
chart of the device operation
[77]

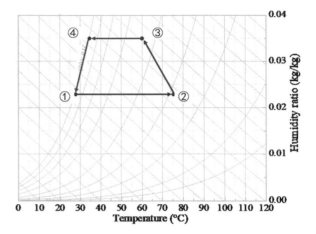

the temperature falls down since the desorption is an endothermic process; the high
humidity air condenses in condenser (point 3) to reach dew point (point 4) and
precipitation water (process 4-1).

5.2.2.2 Experimental Results and Performance Analysis

(1) Experimental results for sorption phase

In the sorption stage, the temperature and relative humidity at the inlet and outlet of
the sorption bed are recorded. The relationship between temperature, humidity and
time is shown in Fig. 5.52. At the beginning of the sorption phase (Fig. 5.52a), the
temperature of the sorbent bed inlet (TSI) is 25 °C, and then TSI changes slowly,

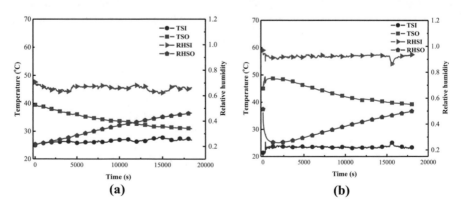

Fig. 5.52 Temperature and relative humidity under two different conditions, **a** 26 °C, 67% RH;
b 23 °C 90% RH [77]

rising from 26 (10,000 s) to 27 (15,000 s). At the beginning of sorption, the sorption proceeds rapidly and the sorption heat is large, so the TSO is as high as 40 °C in the initial stage. Due to the slow sorption reaction rate, TSO gradually decreases to 31 °C. The relative humidity of RHSI at the inlet of sorbent bed fluctuated greatly, ranging from 0.62 to 0.7, which is determined by atmospheric conditions, and RHSO increases steadily with time.

The experiments for 23 °C and 90% humidity is presented in Fig. 5.52b. TSI is remained approximately the same of 23 °C over time and RHSI is constant at 90%. TSO is nearly 49 °C at the beginning of the reaction and turns to 40 °C at the end of the sorption. Over all, the temperatures of beginning and end of experiments in Fig. 5.52b are 9 °C higher than that shown in Fig. 5.52a, and the RHSO shown in Fig. 5.52b is slightly higher than that shown in Fig. 5.52a.

The amount of water vapor in the sorbed phase is calculated as follows:

$$\Delta D = 0.622 \frac{\varphi P_s V_m}{P - \varphi P_s} \tag{5.26}$$

where ΔD is the real-time water vapor volume, kg/s; φ is the relative humidity; P_s is the saturated water vapor pressure, Pa; V_m is the mass flowrate of dry air, kg/s; P is the atmospheric pressure, Pa.

The mass of water vapor calculated from Eq. 5.26 is shown in Fig. 5.53. When the sorption time increases, the outlet water vapor increases, because the sorption rate of sorbent decreases with the reaction. The water vapor at the outlet is affected by the sorption rate and the inlet steam volume. Compared with the data in Fig. 5.53b, the inlet water content in Fig. 5.46a fluctuates significantly, which is consistent with the TSI and RESI phenomena shown in Fig. 5.52a. Although the inlet humidity is basically the same, the outlet water content in Fig. 5.53a is higher than that in Fig. 5.53b, which also means that the sorption performance is worse under the condition of Fig. 5.53a.

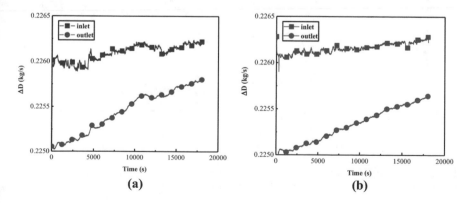

Fig. 5.53 Water vapor amount of inlet and outlet of sorption bed under two different conditions, **a** 26 °C, 67% RH; **b** 23 °C, 90% RH [77]

The water vapor mass sorbed by sorbent bed is calculated by:

$$\Delta d = 1000(\Delta D_{\mathrm{I}} - \Delta D_{\mathrm{O}}) \tag{5.27}$$

The mass of sorbed water vapor is shown in Fig. 5.54a. The total water sorption mass in Fig. 5.54a is 15.2 kg at 23 °C and 90% RH, while it is 13.9 kg at 26 °C and 67% RH shown in Fig. 5.54a, although the water content at the inlet is equal under both conditions. The difference is caused by the difference of sorption potential or free sorption energy ΔF, which is a key concept that combines pressure and temperature as a parameter, and establishes a one-to-one correspondence relationship with water sorption [63, 68, 78].

$$\Delta F = RT_S \ln \frac{P_{sat}}{P_V} \tag{5.28}$$

where P_v is the vapor pressure, P_{sat} is the saturated vapor pressure at the sorption temperature T_s, and R is the gas constant of 8.314, J/(mol K).

In fact, relative humidity (RH) refers to vapor pressure divided by saturated vapor pressure, $RH = P/P_{sat}$, so the points of ΔF can be directly obtained from relative humidity and temperature. As shown in Fig. 5.54b, the ΔF ranges from 50 to 60 kJ/kg for 26 °C and 67% RH, and 8–13 kJ/kg for 23 °C and 90% RH. The higher the value of ΔF, the stronger the ability for sorbing water vapor. According to Eq. 5.28, ΔF is proportional to temperature and inversely proportional to relative humidity. Because the numerical change of temperature has little effect on the thermodynamic temperature, ΔF is mainly proportional to the relative humidity. The results show that 23 °C and 90% RH have a good sorption capacity for higher relative humidity.

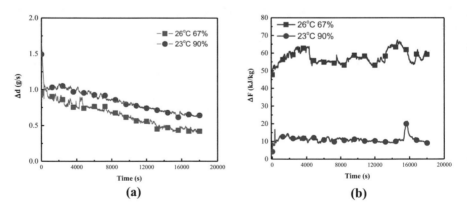

Fig. 5.54 The sorption performances under two different conditions, **a** the water vapor mass sorbed by sorbent bed; **b** change of free energy ΔF during the sorption process [77]

(2) Experiment result in desorption phases

After 1800 s, the desorbed vapor occurs at the desorption stage of the heated sorbent bed which is sufficiently heated, and TSI increases rapidly during 0–1800 s. During 1800–3750 s, TSI decreases slightly from 75 to 70 °C due to condensation, and then remains at about 77 °C until the reaction is completed, as shown in Fig. 5.55a.

At the same time, TSO increases rapidly during 0–1800 s and the temperature difference between TSI and TSO is about 20 °C, which is mainly due to the endothermic reaction in the desorption process. TSO decreases with the increase of TSI, then it remains at 57 °C between 5000 and 15,000 s, and then increases with the completion of desorption process.

The RHSI decreases from 0.5 to 0.2 because the absolute water content of the heater remains constant between 0 and 2500 s and then it keeps at 0.2. RHSO increases significantly from 0.55 to 0.72 in the 0–1800 s, decreases to 0.67 in the 1800–3750 s. Then it remains at about 0.7 during the period of 50,000–150,000 s, and it shows a stable downward trend after 150,000 s.

As shown in Fig. 5.55b, the condenser inlet temperature (TCI) changes with the change of TSO in the whole process, but the value of TCI is slightly less than that of TSO. For example, since the humid air has been precooled in the cover area, TCI is 52 °C and TSO is 57 °C at 1800 s. Similarly, the relative humidity at the condenser inlet (RHCI) varies with RHSI and it is slightly higher than that of RHSI.

At the same time, when condensation doesn't happen, the outlet temperature (TCO) of condenser increases rapidly in 0–1800 s, which is the same as TCI. Then it drops sharply from 52 to 32 °C in 1800–3750 s; then it keeps at 37 °C from 3750 to 25,000 s.

The relative humidity at the outlet of condenser (RHCO) increases with the increase of RHCI during 0–1800 s, and reaches the peak value of 0.92 at 1800 s. After that it remains at about 0.9 during 5000–15,000 s, and it also decreases steadily after 15,000 s.

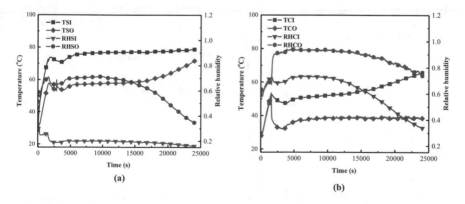

Fig. 5.55 The desorption performance under 77 °C, **a** temperature and relative humidity data of sorbent bed; **b** temperature and relative humidity data of condenser [77]

The total operation time of desorption reaction is 24,000 s. Because the RHCO is less than 0.75 after 24,000 s, which means that the partial pressure of water vapor for the circulating liquid is low, and the wet air can't reach the dew point, so the desorption is completed and no more water can be collected.

The desorption performance of sorbent bed and condensation behavior of condenser are shown in Fig. 5.56. For the sorbent bed, the desorption steam mass is obtained by the moisture mass difference between the outlet and the inlet of the sorbent bed. For the condenser, the quality of condensate is obtained by the water difference between the inlet and outlet of the condenser. For the sorbent bed, water is released from the sorbent at the beginning, rising rapidly to 0.8 g/s at 1800 s, and then rapidly decreasing to 0.45 g/s at 2600 s since the condensation proceeds. Then the value increases steadily from 0.45 to 0.7 g/s during 4000–5100 s. It maintains at 0.7 g/s before 10000 s, and then falls to 0.6 g/s at 20,000 s. For condenser, the water begins to precipitate out after 1800 s, then it is reduced from 0.72 to 0.45 g/s during 1800–3750 s. Then it remains around at 0.5 g/s, which is smaller than that of sorbent bed, i.e. 0.7 g/s.

(3) **The overall performance of the device under the condition of different inlet RH and desorption temperature**

The overall performance of the device is tested under the relative humidity of 85, 75 and 65%. Taking 75% as an example, the sorption phase is treated at 25 °C and 75% RH for 5 h, and then desorbed at 77 °C inlet temperature for 5.5 h. Finally, 14.7, 13.6 and 12.5 kg fresh water are obtained by condensation at 85, 75 and 65% relative humidity. The higher the inlet value of relative humidity is, the more water is obtained. Under the same sorption and desorption time, the inlet relative humidity is 75%. At 90 °C and 60 °C, the collected water is 14.5 kg and 0 kg, respectively. Comparing the data at 77 and 90 °C, the higher the temperature is, the more water is collected.

Fig. 5.56 The desorption and condensation performance at 77 °C [77]

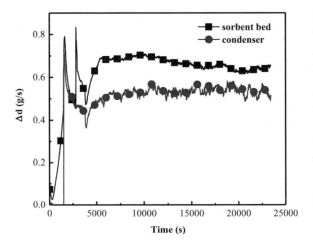

5.2.3 The Test System Driven by Solar Power

5.2.3.1 Introduction of a Semi-open Solar Driven Device

As shown in Fig. 5.57a, the improved semi-open device consists of four parts: solar air collector, sorption bed, fan and condenser. Solar air collector is used to provide hot air for heating and desorption phase and the sorption bed adopts corrugated plate structure. Compared with the conventional stacking method, this design has the advantages of compact structure and high efficiency. The sorbent is directly used to form a mass transfer channel to promote air circulation and make the sorption and desorption phase sufficient. Environmental air driven by fans is utilized for cooling and sorption phase and water cooled condenser is utilized. The photos of the freshwater production appliance from different sides are shown in Fig. 5.57b. The total size of the device is $1.2 \times 1.0 \times 1.3$ m with $0.4 \times 0.4 \times 0.6$ m size of sorbent bed.

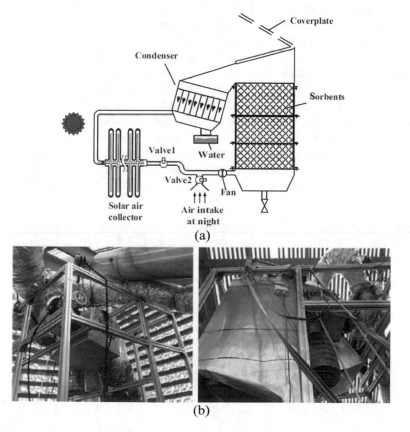

Fig. 5.57 The solar powered device. **a** Schematic diagram of the fresh water production system; **b** real photos of the improved device [79]

Actually, the system is open-type in sorption and closed-type in desorption as shown in Fig. 5.57a. The working processes are as follows:

(1) Sorption process: the cover plate (Fig. 5.57a) is open, valve 1 is closed and valve 2 is open. The centrifugal fan is open and ambient air flows into the equipment and passes through the sorbent from the bottom of the bed. The water vapor in the air is sorbed by the sorbent and the air then flows through the cover plate to the environment.

(2) Desorption process: cover plate is closed (Fig. 5.57a), valve 1 is open and valve 2 is closed. The air from the condenser is heated in a solar collector and then flows to the sorption bed. The bed is heated by hot air and the water vapor is desorbed. After that, high relative humidity air flows into the condenser, where the water vapor condenses into liquid water, which is collected by gravity at the bottom of the condenser.

Figure 5.58 shows the psychrometric chart, with four main points in the cycle. Point 1 is the initial condition of air in the solar collector, and the air is heated with constant absolute water content in the solar collector, as described in process 1–2; compared with concept machine [79], the desorption process of 2–3 is more intense due to the large temperature difference between point 2 and point 3; moreover, the relative humidity of point 3 is 80%, which is easier to reach dew point (point 4) and precipitate more water than concept machine (process 4-1). The saturated air is then recycled and re-enters the solar collector. The difference of humidity ratio between process 1–2 and process 3–4 is 0.043 kg/kg, which is half of the concept machine, but the air flow of solar driven device is much larger than that of concept machine.

The structure of vacuum tube air collector is shown in Fig. 5.59a. The saturated air from the condenser enters several branches of the main air duct through external force, and the vacuum tube directly heats the air by solar radiation. After that, high temperature air is discharged from the pipe to desorb the sorbent in the bed. Four

Fig. 5.58 Psychrometric chart of the solar driven device [79]

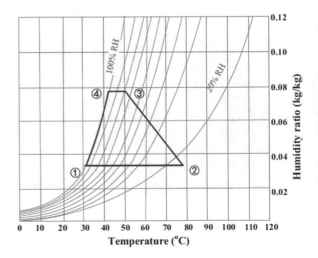

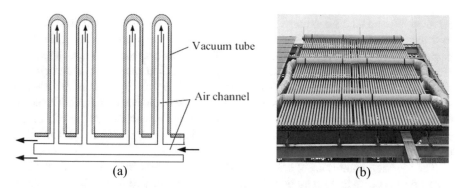

Fig. 5.59 **a** Structure of evacuated tubular air collector; **b** solar collector adopted in system [79]

groups of vacuum tube air collectors constitute the whole heat collection system of fresh water production plant. The total area of each group of collectors is 3 m², and the vacuum tubes are arranged in parallel with an inner diameter of 47 mm and an outer diameter of 58 mm, as shown in Fig. 5.59b. Compared with solar collector, the main advantages of air circulation are low material requirements, no corrosion, simple system, low price, small volume and heat loss. The disadvantage is that the specific heat capacity of air is far less than that of water.

5.2.3.2 Experiment Results of the Semi-open Solar Driven Device

The sorption performance of the solar driven device is shown in Fig. 5.60. During the sorption process, the inlet and outlet air conditions of the sorbent bed are shown in Fig. 5.60a. The inlet temperature (TSI) and the relative humidity (RHSI) of the sorbent bed inlet fluctuate greatly from 0 to 8000 s. After that TSI keeps at 30 °C, the RHSI is remains among 85% between 8000 and 30,000 s. Then, TSI increases from

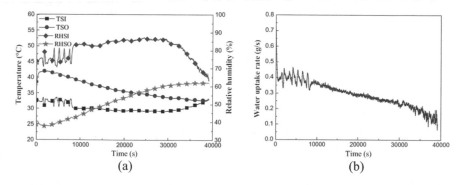

Fig. 5.60 Sorption performance of the solar driven air-to-water device in 20/07/2016. **a** Temperature and relative humidity of sorption bed; **b** sorbed water mass in sorbent bed [79]

30 to 33 °C, and RHSI decreases from 85 to 65% between 30,000 and 40,000 s, due to the release of sorption heat, the temperature of TSO at the outlet of sorbent bed begins to be 43 °C, and then decreases to 33 °C with the decrease of sorption reaction rate. At the same time, the relative humidity of RHSO at the outlet of sorbent bed increases steadily from 38 to 62% with time. As shown in Fig. 5.60b, the mass of water sorbed in the sorption bed is the difference between the inlet and outlet steam mass obtained from Eq. 5.26. The initial mass fluctuates with the change of TSI and RHSI, and then decreases from 0.35 to 0.15 g/s between 10,000 and 38,000 s. The total sorption capacity is 11.2 kg.

The working conditions of solar air collector are shown in Table 5.9. During the desorption process of sorbent bed, the velocity of hot air should be slow in order to realize the full desorption reaction between sorbent and air, so the low-power fan is used.

The performance parameters of the solar air collector system are shown under summer condition in Fig. 5.61. Because the desorption stage of the unit is a closed cycle, the inlet gas temperature of the collector is stable between 45 and 47 °C in Fig. 5.61a. At the same time, the real-time solar radiation intensity gradually increases from 700 to 930 W/m² from 9:00 to 12:00, and drops to 375 W/m² at 17:00. In Fig. 5.61b, the gas at the outlet of the solar collector is basically constant at 80 °C between 9:00 and 13:00. After 13:00, the temperature gradually decreases from 80 to 55 °C. The thermal efficiency of the solar air collector is derived from the following Equation

Table 5.9 Operating condition of the solar air collector driven system [79]	Parameters	Value
	Collector area	12 m²
	Air volume flow	64 m³/h
	Power of fan	50 W

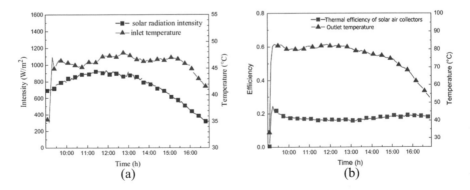

Fig. 5.61 a Temperature of collector inlet air and the solar radiant intensity in 20/07/2016; **b** temperature of outlet and the thermal efficiency of solar air collector in 20/07/2016 [79]

$$\eta = Q_g/Q_{solar} = \frac{C_{pg}m_g(t_{so} - t_{si})}{IA_{solar}} \quad (5.29)$$

In fact, the solar collector is used as the auxiliary equipment of fresh water production to provide heat for the desorption stage. The average thermal efficiency is 0.2, which is far lower than 0.66 for other solar air collectors [79]. Therefore, under the same condition as the high efficiency collector, the collector area can be reduced from 12 to 4 m^2. Otherwise, the thermal efficiency in Fig. 5.61b is constant at 19%.

During the desorption process, the sorbent bed is fully heated by the hot air of the solar collector from 9:00 to 9:20, and a large amount of water vapor is released from the sorbent bed and condenses. As shown in Fig. 5.62a, TSI rises rapidly at the beginning of the reaction, and then decreases from 87 to 83 °C under condensation starting from 9:20. After that, the temperature is maintained at 84 °C from 10:00 to 13:00, and then decreases to 57 °C at 16:30 until the reaction is completed. At the same time, because the desorption process is an endothermic reaction, TSO increases rapidly at about 28 °C. Then TSO remains stable at 58 °C from 10:00 to 16:30 until the desorption process is completed. Due to the absolute water content of solar collector remains unchanged in the initial stage, RHSI decreases from 80 to 20%, and then increases with the decrease of TSI. After that it keeps at about 23% from 9:30 to 13:00, and then gradually increases to 40% from 13:00 to 16:30. When a large amount of water vapor is desorbed from the sorption bed, the RHSO increases sharply from 60 to 82% in the first 20 min, and then it stays about 80% as a constant. Between 14:00 and 16:30, it shows a steady decline to 50%. As shown in Fig. 5.62b, the condenser inlet temperature (TCI) varies with TSO, which is 6 °C lower than TSO because the humid air has been precooled during the process from the sorbent bed to the condenser. For example, TCI is 53 °C (Fig. 5.62b), while TSO is 59 °C at 9:30 (Fig. 5.62a). Similarly, the relative humidity at the condenser inlet (RHCI) varies with RHSI and is about 8% higher than that of RHSI in the whole reaction process.

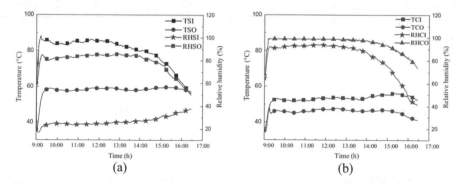

Fig. 5.62 Desorption performance of the solar driven device in 20/07/2016. **a** Temperatures and relative humidity of sorbent bed; **b** temperatures and relative humidity of condenser [79]

Fig. 5.63 Condensed water
mass of the solar device in
20/07/2016 [79]

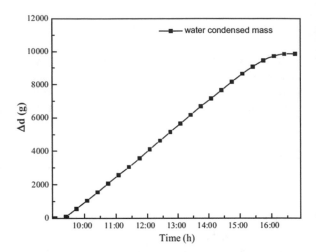

Fig. 5.63 Condensed water mass of the solar device in 20/07/2016 [79]

At the same time, the condenser outlet temperature (TCO) increases from 35 to 47 °C in 30 min. Because there is no condensation, TCO value is the same as TCI. Then, the temperature drops from 48 to 46 °C from 9:00 to 15:00. After 15:00, the temperature drops slowly to 40 °C at 16:30. The relative humidity at the outlet of the condenser (RHCO) increases sharply from 65 to 93% with RHCI at 30 min. Then, it keeps at about 90% from 9:30 to 14:00, and shows a stable downward trend after 14:00. The total operation time of desorption reaction is 7.5 h, since TSI is lower than 60 °C after 16:00, which is too low to desorb the sorbent. Because of the low partial pressure of water vapor in the circulating liquid, the dew point could not be reached, so the desorption is completed, and the water could not be collected after 16:00.

The condensate quality is shown in Fig. 5.63. After 9:30, 10.3 kg water is condensed from 9:30 to 16:00. After 16:00, a large amount of water condenses to 0, indicating that the desorption process is basically completed, and the total water volume remains at about 10.3 kg. The calculated mass of water is 10.3 kg, while the actual collected water mass is 9 kg because a part of saturated humid air from the condenser condenses in the pipe.

5.2.4   Scalable Prototype with Energy Storage Tank

5.2.4.1   Prototype Fabrication

The size of sorber layer is 1.0 × 0.75 × 0.3 m, and it can hold 70 kg of sorbent. The thickness of the ACF felt is 5 mm. The enclosure of sorption bed is prepared by welding 304 stainless steel plate (2 mm thick), then corrosion-resistant epoxy boards (5 mm) is attached on all side walls inside to separate the sorbent from the stainless steel plate. The bottom face of the enclosure is welded with 304 stainless steel orifice plate (1.5 mm thick, 6 mm hole diameter and 8 mm hole spacing), and the moist/hot air will go through the holes of steel plate to enter sorbents during sorption/desorption processes. A large number of sorbents are processed into a plate or corrugated structure and then integrate into this enclosure by placing layer and layer vertically to form the honeycomb-structure sorbent bed. Two layers of such sorbent bed are processed.

Electric heater is manufactured to simulate the solar air collectors with $1.3 \times 0.4 \times 0.4$ m size. The heater box is fabricated by welding stainless steel with 8 kW maximum power of electric heating tube inside. A precise temperature control box is set to control the heating power of the tube. A fan is connected to a frequency converter with the 200 W rated power, and the air volume in sorption and desorption processes is 457 m^3/h. The mixture, 4:1 mass ratio of stearic acid and expanded graphite, is pressed into the block (880 kg/m^3 packing density and 15 W/(m K) average thermal conductivity). This tank is used to store a part of energy by phase change of stearic acid to make the inlet temperature of sorbent bed stable. 40 copper tubes (25 mm diameter) are adopted as air flow passages in the storage cylinder, and high thermal conductivity of copper can quickly transfer heat between air and mixture materials. A plate-fin water cooling condenser ($510 \times 530 \times 165$ mm size, 20 m^2 heat transfer area) is adopted for higher RR level, and the condensing temperature can be adjusted by changing the water flow. According to the operation condition, we set the condenser average temperature around 30–33 °C, which is the air temperature of the arid/desert area at daytime. T_c is the condenser temperature, T_{in} is the inlet water temperature, T_{out} is the outlet water temperature. During the experiments, the water inlet temperature is 18 °C, the water outlet temperature is 45 °C, so the condenser temperature is 30 °C.

$$T_c = \frac{(T_{in} - T_{out})}{\ln(T_{in}/T_{out})} \tag{5.30}$$

A fan is adopted to take in a large amount of air (this could be replaced by solar chimney combined with solar air collector), while the air flow rate should be slow for the sufficient reaction between water vapor and sorbents in sorption or desorption process. The plate and corrugated sorbents are placed layer by layer vertically, in which the air flows from the bottom to the top of the sorption bed during sorption and desorption phases. This design greatly improves the heat and mass transfer performance.

As shown in Fig. 5.64, the water harvesting production device can be divided into five main sections: sorption bed, heat storage tank, electrical heater (simulating solar air collector), fan and condenser. The unique design of sorbent bed is the most important part.

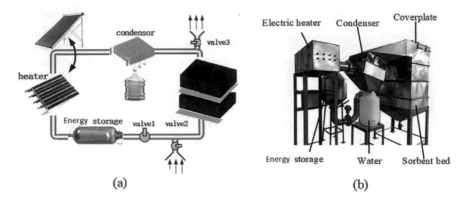

(a)   (b)

Fig. 5.64 Illustration of the sorbent structure and the schematic of water harvesting device, **a** schematic of the device (valve 1 is closed, valve 2 and valve 3/cover plate are open in sorption process. Valve 1 is open, valve 2 and valve 3/cover plate are closed in desorption process); **b** real device [80]

5.2.4.2   Performance Analysis

The water harvesting devices is designed and manufactured by filling the solidified composite ACF-LiCl sorbent [79] with the optimized structural design in sorbent bed. It can output 38.5 kg water per day under the atmosphere condition of 25 °C/75% RH in sorption process, and 60 °C sorbent temperature/26.9 °C condensation temperature in desorption process.

In order to study the water harvesting cycle of the device, the air sorption conditions under humid and dry climate conditions is selected, which are 25 °C/75% RH and 25 °C/39% RH, respectively. The test results are presented in Fig. 5.65. In the desorption stage, I, O, S, C represent the inlet, outlet, sorbent and condenser respectively, and TS is the average temperature of sorbent. In the desorption stage, the condensation occurs after 100 min when the sorption bed is heated enough to desorb the water vapor.

Assuming that the duration of sorption and desorption is same, the moisture recovery (RR) of the sorbent bed can be calculated by the following formula:

$$RR = \frac{M_{water}}{\rho_{air,sorber} Q_{air,sorber} t_{sorber} d_{sorber,i}} \approx \frac{d_{desorber,o} - d_{cond}}{d_{sorber,i}} \qquad (5.31)$$

As shown in Fig. 5.66, the experiment dynamic processes of air states change with time. Point a1 and a2 (Fig. 5.66a) are the inlet and outlet atmosphere condition. The outlet air temperature rises up to point a2′, which is the cross point of constant enthalpy line and constant relative humidity line. Then the sorption will turn to a constant enthalpy process from a2 to a1. In desorption process, point d1 and d2 are the conditions for inlet and outlet air, which varies from d2 to d1 along with a constant enthalpy process. Then the outlet moist air from the sorber will be condensed

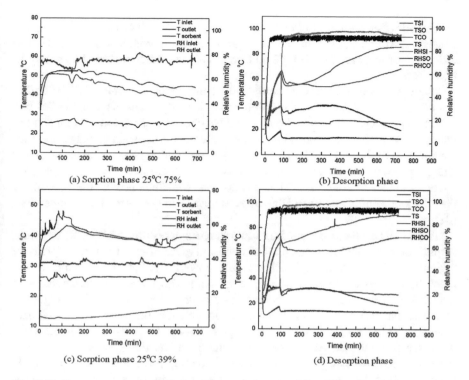

Fig. 5.65 Temperature and relative humidity record under 25 °C, 75% RH and 25 °C 39% RH conditions. **a** Sorption phase under 25 °C, 75% RH; **b** desorption phase after 25 °C, 75% RH sorption; **c** sorption phase under 25 °C, 39% RH; **d** desorption phase after 25 °C, 39% RH sorption [80]

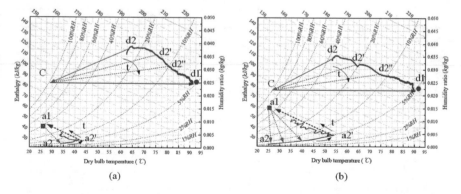

Fig. 5.66 The water harvesting cycles under different conditions. **a** Arid climate; **b** humid climate. The inlet air conditions for sorption process are: **a** 25 °C, 39%; **b** 25 °C, 75%. C-to-d1 presents the heating process, d1-to-d2 presents the desorption process, d2-to-C presents the condensing process. a1-to-a2 presents the sorption process. The inlet air conditions for desorption process are 92 °C, 5% [80]

into liquid water, from d2 to Point C, which is the dew point of the air leaving the condenser. After that, the saturated air is heated in a solar collector till point d1. The difference of moisture between d2 and point C is the water collected as shown in Fig. 5.66b.

Globally, whether in arid or humid areas, the outlet air state of AWH sorber can be described as follows. In sorption phase, it firstly changes along the RH line (Fig. 5.66), which is determined by the sorbent characteristics, and the temperature increases. After that the temperature decreases along the constant enthalpy line with the direction of the sorber inlet air state (a1). In desorption phase, it changes almost along the relative humidity line near the end of the sorption state, and the temperature increases gradually. Afterwards along the constant enthalpy line with the direction of the inlet temperature of sorber. Ultimately, it reaches to the sorber inlet temperature (d1).

Table 5.10 lists the corresponding water harvesting results. Additionally, another sorption climate condition is increased to apply in exceedingly dry climate condition regions (as Yazd at April night). The result shows the higher the humidity is, the more the water is collected, especially considering that the atmosphere conditions fluctuate with the seasons. Water harvesting collected is obtained by weighting. The water harvested is 14.3 kg water, even though under the extremely dry climate condition. The recovery ratios are very high in a range of 41–54% and it generally decreases with the RH. According to existing literature, this system is the best in records. Water harvesting efficiency (water harvested/sorbed) reaches up to 70%, which indicates the compactness and effectiveness of the system. An electric power

Table 5.10 The experimental performance of the device under different inlet air conditions [80]

Conditions	Performance		
	Test 1	Test 2	Test 3
Sorption			
T_{ads} (°C)	25	25	15
RH_{ads} (%)	75	39	37
Total water in the imported air	94.9	46.7	26.5
Water sorbed	63.8	29.8	19.6
Desorption			
T_{air} (°C)	92	92.8	89.3
T_{cond} (°C)	29	32.2	33
T_{des} (°C)	68	72	74
Water-harvesting collected (kg)	38.5	20.7	14.3
RR (%)	41	44	54
Harvesting coefficient (Hc) (%)	60	69	73

Test 1: moist climate condition (simulating the climate of Garissa, Kenya or offshore area); Test 2: normal climate condition in desert (simulating the climate of Kharga, Egypt, July night); Test 3: extremely dry climate condition in desert (simulating the climate of Yazd, Iran, April night)

Table 5.11 The water harvest prediction in different areas around the world [80]

	Kharga Egypt arid	Yazd Iran arid	Oodnadatta Australia arid	Minfeng China arid	Garissa Kenya semi-arid	Trivandrum India offshore
Water harvest prediction Ton/year	8.23	4.76	8.47	6.19	12.40	18.53
Solar collector area needed/m^2	20.5	14.6	22.7	23.3	24.3	36.4

meter is employed to show the power consumption of the device. It costs 2.3 kWh of heat to get 1 kg freshwater under 25 °C/75%(RH) condition, while it costs 2.7 kWh of heat under 25 °C/39%(RH) condition.

It is worthy noted that the inlet air temperature of desorption can be easily achieved in the most area according to the solar air collector performance [80]. For instance, the outlet air of solar collector can reach 90 °C (108 °C) under 400 W/m^2 (600 W/m^2) solar radiation intensity with the 66% average thermal efficiency of solar air collectors [81]. Sun radiation has an intensity of 20 MJ/m^2 per day if a solar air collector has an efficiency of 70%. Then 1 m^2 solar collector could yield 2 kg water per day.

As shown in Table 5.11, the critical inherent parameters of AWH device under different working conditions are calculated, which is RR and harvesting coefficient. According to these parameters, the water harvesting mass potential in many various regions, and the local humidity temperature and solar irradiation intensity can be reckoned. Table 5.11 also shows the water harvest potentials in different areas around the world. The device can harvest 20–30 kg water with 15 m^2 area of solar collector owing to the strong radiation intensity in desert area. The moist atmosphere area can collect more water (Garissa or Trivandrum). The total daily water requirement for an adult is 5 kg. As a result, the water need of 5 adults can be meet by one water harvesting device in desert area. Such a device can also provide the water for crops to increase the food production by increasing adequate freshwater [82].

5.3 Solid Sorption Cycle for Eliminating NO$_x$ Emission

Diesel engine exhaust purification technology includes diesel oxidation catalyst (DOC), diesel particulate filter (DPF) and de-NO$_x$ purification catalyst (De-NO$_x$). DOC technology is mainly used to treat HC, CO and soluble organic (SOF). DPF technology is used to treat small particles in exhaust gas. De-NO$_x$ technology, as the name suggests, is used to eliminate nitrogen oxides. More and more attention has been paid to the De-NO$_x$ technology of NO$_x$ emissions from diesel vehicles. From Euro I (8.0 g/km) in 1992 to Euro VI (0.5 g/km) in 2012, European requirements for diesel vehicle NO_x emission standards have become more and more stringent. As the country with the most stringent diesel vehicle emissions, the United States implemented a NO_x emission limit of 0.8 g/km in 2007, which is far lower than the 2008 Euro V standard (2.0 g/km).

With reference to diesel engine NO_x purification technology, the existing and mature technical solutions mainly include ammonia selective catalytic reduction of NO_x (NH_3-SCR), hydrocarbon selective catalytic reduction of NO_x (HC-SCR), storage reduction technology (NSR), NO_x and PM purification technology and four-way catalytic technology [83]. At present, the de-NO_x technology widely used in the world's top automobile manufacturers is urea SCR technology, namely NH_3-SCR technology.

For liquid materials used as ammonia sources, urea solution is used instead of ammonia solution, which is selected to reduce NO_x for many years. The urea solution is decomposed into NH_3 in the exhaust pipe under the heating of the exhaust gas, and completely reacts with NO_x in the catalytic converter, and is converted into N_2 and H2O. According to the latest data provided by Bosch, an 80% NO_x conversion rate requires a maximum volume flow of 50 ml/min.

However, urea SCR technology has many shortcomings in practical applications, such as low NO_x conversion efficiency at low temperature, the high cost, incomplete decomposition and coking, urea crystallization, and low effective ammonia content (the theoretical ammonia content of 18.42% when urea solution is about 32.5%). Therefore, the development of a new ammonia storage technology with better performance is of great significance.

Some researchers are working hard to find new NH_3-SCR technology. Cottle [84] first proposed the idea of using dissolved salts to store ammonia in 1959, and designed a storage-retrieving ammonia recovery system. In 1992, Walker and Speronello [85] summarized the design and development of ammonia injection system and proposed that teflon and stainless steel can be used as raw materials for ammonia injection. Chen and Lin [86] designed a calculation model to determine the thermal response of an ammonia plant on fire in 1998. It is good to avoid the safety problems of ammonia storage. Since entering the twenty-first century, solid sorption ammonia storage technology has gradually received attention. In 2007, Kaboord [87] studied the cylindrical honeycomb sorption bed with molecular sieve as the sorbent, and proposed a composite sorption ammonia storage and regeneration system composed of two ammonia containers. The $CaCl_2$-$CaBr_2$ mixture and $MgCl_2$ sorbent were studied for ammonia storage technology by Chun and Ken-ichi in 2004 [88] and Elmøe et al. in 2006. Later in 2008, Johannessen et al. [89] found that the storage capacity of magnesium-based complexes and calcium-based complexes to release ammonia was approximately three times that of urea. Veselovskaya and Tokarev [90] proposed a new ammonia $BaCl_2$/vermiculite composite sorbent, and studied the quasi-isobaric ammonia sorption/desorption kinetics on this new composite sorbent in 2011. In 2009, a series of experiments were conducted by Fulks et al. [91] to investigate the performance of different ammonia sources, including 3 types of urea,

9 types of ammonia salts, 9 types of metal ammine and 2 types of baselines (anhydrous liquid ammonia and acetone). It was found that ammonium salts and the metal ammonium chloride has the best performance.

The development of solid materials as ammonia sources is a relatively stable method. For chemical solid sorbents, the halide always has attenuation. In order to solve this problem, the composite sorbent has been studied, and the results show that the heat and mass transfer performance of the composite sorbent has been significantly improved. Wang et al. [92] studied composite sorbents that it improved heat and mass transfer performance while increasing sorption capacity, and summarized the development methods of composite sorbents in detail. Literature [93, 94] conducted experimental research on expanded natural graphite (ENG) and expanded natural graphite treated with sulfuric acid (ENG-TSA) in composite sorbents. The results show that ENG-TSA matrix has higher thermal conductivity and reasonable permeability. The experiments and researches in the literature [95, 96] have made great efforts to improve the sorption capacity and sorption rate. However, there are more researches on the application of composite chemical sorbents (CCS) in refrigeration and energy storage, but less research on the treatment of NO$_x$ emissions.

This section proposes a composite sorbent SCR. The ammonia gas releases from a solid ammonia source and directly injects it into the exhaust pipe. The effects of the composite chemical sorbent (with ENG-TSA as matrix)-NH$_3$ on the treatment of NO$_x$ emissions are analyzed and compared.

5.3.1 Working Principle

The ASS device is installed in the position where the AdBlue box and the AdBlue pump are installed under the truck underframe. The ASS unit is close to the exhaust pipe, which can more conveniently use the waste heat.

For the NOx conversion process, the composite sorbent SCR system (Fig. 5.67a) consists of an ASS unit, a SCR catalytic converter (SCR-CC), a SCR electronic control unit (SCR-ECU), a nozzle, valves, several sensors, several connecting lines and some pipes. In more detail, the SCR catalytic converter mainly includes a carrier (usually shaped as a honeycomb structure), a catalyst and encapsulation. The sensors mainly include pressure sensors, temperature sensors, NH$_3$ sensors and NO$_x$ sensors. For safety reasons, a pipe branch is designed, and two electromagnetic valves are used to control the exhaust flow (indirectly control the reaction temperature). There are mainly two working stages: NO$_x$ conversion and NH$_3$ sorption.

(1) Desorption and de-NO$_x$ stage (Fig. 5.67a). The ASS device is driven by electric heater and exhaust gas. Before starting the vehicle, preheat the system with electricity for quick injection. After the vehicle is started, the heat source is switched to exhaust gas to save energy. During the initial heating stage, the temperature and pressure in the ASS device will increase. When the pressure in the device is higher than the equilibrium reaction pressure, ammonia is desorbed.

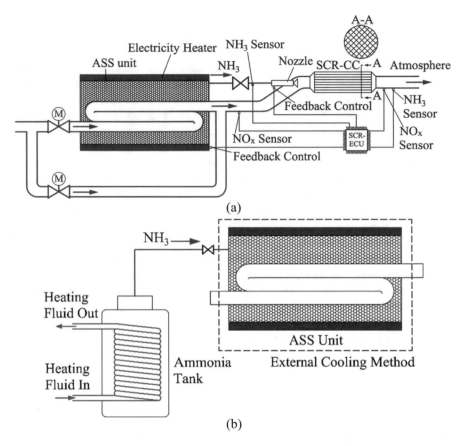

Fig. 5.67 Working principle of the system. **a** Desorption and de-NO$_x$ phase; **b** NH$_3$ sorption and storage phase [97]

After that, the NH$_3$ gas is ejected through the nozzle, flows into the exhaust pipe, and mixes with the NO$_x$ in the exhaust gas. The mixed gas reacts thoroughly in SCR-CC, and the reaction product passes through the NO$_x$ sensor, which returns the feedback signal to SCR-ECU. After that, the SCR-ECU adjusts the NH$_3$ injection amount based on the NO$_x$ content in the final exhaust. In SCR-CC, the reactions of eliminating gaseous pollutant of NO$_x$, CO, CO$_2$ and other particulate matter (PM) pollutants are catalyzed.

(2) Sorption and storage stage (Fig. 5.67b). After all ammonia stored in the ASS device is used up, NH$_3$ sorption and storage processes are required. In this process, the ASS device is connected to the ammonia tank. The ASS device can be cooled by external cooling, and the pressure and temperature are reduced. The ammonia tank is heated by ambient medium, and its temperature is at least 5 °C lower than the sorption temperature in the ASS device to avoid condensation in the device. When the internal pressure of the ASS device is lower than the

pressure of the ammonia tank, the valve opens and ammonia is sorbed into the ASS device until it reaches saturation.

Since the ASS device requires rapid sorption storage and rapid ammonia release process, the heat and mass transfer performance are the key factor of the ASS device. Studies have shown that adding ENG-TSA can significantly improve the heat transfer performance without affecting mass transfer [98]. Composite sorbents are also a solution for long-term performance degradation.

5.3.2 Theoretical Performance Analysis of Halide-NH$_3$ Chemisorption

For the NH$_3$ injection process, the pressure of the ASS device is usually 2 bar higher than the pressure at the nozzle outlet, which is 2 bar. This means that the pressure inside the ASS device should be around 4–5 bar. For the heat exchanging process between the exhaust gas and the ASS device, the temperature of the device is generally lower than 250 °C due to the low heat transfer coefficient of the gas. Considering the safety issues, the desorption temperature is controlled at 50–100 °C through the design of pipe branches. At the same time, considering that better sorption performance and faster sorption rate can be obtained under lower cooling temperature conditions, the sorption temperature should be adjusted according to seasonal changes to obtain the best cooling effect.

Two typical diesel vehicles are analyzed. One is a small car (Type A) with a 2 L engine displacement, and its fuel consumption per 100 km is about 7 L; the other is a large car (Type B) with an engine displacement of 7 L, and its fuel consumption per 100 km is about 35 L. For Type A trucks, the required NH$_3$ is 30–50 g per 100 km, calculated based on the relationship between NO$_x$ emissions and diesel consumption. Similarly, for Type B trucks, the required NH$_3$ is 150–250 g per 100 km. The sorption quantity required to store NH$_3$ can be calculated through the circulating sorption capacity, and the equation is:

$$m_{ad} = \frac{m_{am}}{(x_{ad} - x_{des})} \tag{5.32}$$

Different halides have different sorption capacities. For different halides, the ideal condition is the equilibrium condition without sorption hysteresis, that is, the chemisorption process and the desorption process are completely reversible. The heat transfer coefficient of the exhaust gas is very low, so the heat transfer performance between the exhaust gas and the ASS device is very important. The temperature of the ASS device is generally lower than 200 °C. High temperature sorbents (HTSs) require a higher heat source temperature, generally higher than 200 °C [99]. At the same time, it also requires a higher desorption heat, which is the reason for the high desorption enthalpy. Therefore, this is not the best choice. In this study, low temperature sorbents such as NH_4Cl, NaBr, and $BaCl_2$ and middle temperature sorbents such as $CaCl_2$ and $SrCl_2$ were selected. The Equations for LTS are as follows [100]:

$$NH_4Cl \cdot 3NH_3 + \Delta H_1 \leftrightarrow NH_4Cl + 3NH_3 \tag{5.33}$$

$$NaBr \cdot 5.25NH_3 + \Delta H_3 \leftrightarrow NaBr + 5.25NH_3 \tag{5.34}$$

$$BaCl_2 \cdot 8NH_3 + \Delta H_2 \leftrightarrow BaCl_2 + 8NH_3 \tag{5.35}$$

The reaction of LTS-NH_3 complex is one step, whereas the reaction of MTS-NH_3 complex is multi-step. The Equations of MTS are as follows [99]:

$$CaCl_2 \cdot 8NH_3 + \Delta H_4 \leftrightarrow CaCl_2 \cdot 4NH_3 + 4NH_3 \tag{5.36}$$

$$CaCl_2 \cdot 4NH_3 + \Delta H_5 \leftrightarrow CaCl_2 \cdot 2NH_3 + 2NH_3 \tag{5.37}$$

$$CaCl_2 \cdot 2NH_3 + \Delta H_6 \leftrightarrow CaCl_2 + 2NH_3 \tag{5.38}$$

$$SrCl_2 \cdot 8NH_3 + \Delta H_7 \leftrightarrow SrCl_2 \cdot NH_3 + 7NH_3 \tag{5.39}$$

$$SrCl_2 \cdot NH_3 + \Delta H_8 \leftrightarrow SrCl_2 + NH_3 \tag{5.40}$$

Equations 5.38 and 5.40 require high temperature (about 250 °C) at a saturation pressure of 4 bar.

The Clapeyron equilibrium curves of different metal halides-ammonia working pairs are presented in Fig. 5.68.

The ammonia storage capacity of single halide and the weight of sorbents are shown in Table 5.12. The desorption pressure is 4 bar, and the corresponding desorption temperature is shown in Table 5.12 as well. The chemisorption temperature is determined by the ambient temperature, that is, the ASS unit exchanges heat with the heat sink at the ambient temperature. According to different seasons, the temperature varies between 50 and 100 °C and the corresponding chemisorption pressure is shown in Fig. 5.68.

For LTS, the performance of NH_4Cl is better than NaBr, and both have very small weight, i.e., only 31.5–52.4 g per 100 km and 34.6–57.6 g per 100 km. The weight of $BaCl_2$ is much higher than the other two LTSs. For MTSs, $CaCl_2$ and $SrCl_2$ have similar properties, and the weight of MTSs is larger than that of LTSs.

Fig. 5.68 Clapeyron
diagram of metal
halide-ammonia working
pairs [97]

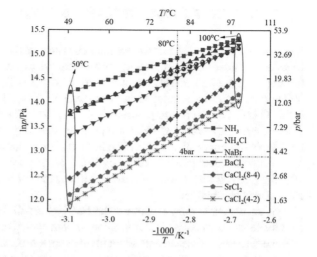

Table 5.12 The weight of chemical sorbent per 100 km [97]

Metal halides	Chemisorption temperature (°C)	Chemisorption pressure (MPa)	Desorption temperature (°C)	Chemical sorbent of type-A (g)	Chemical sorbent of type-B (g)
NH₄Cl	0	0.223	21.0	31.5–52.4	157.3–262.2
	10	0.316			
	20	0.411			
	30	0.592			
NaBr	0	0141	26.8	34.6–57.6	172.9–288.2
	10	0.228			
	20	0.324			
	30	0.440			
BaCl₂	0	0.063	42.0	45.9–76.6	229.7–382.8
	10	0.105			
	20	0.145			
	30	0.252			
CaCl₂(8–2)	0	0.018	60.0–72.0	42.6–71.1	213.2–355.4
	10	0.038			
	20	0.058			
	30	0.104			
SrCl₂(8–1)	0	0.015	66.7	44.3–73.8	221.3–368.8
	10	0.026			
	20	0.045			
	30	0.069			

5.3.3 Results of Composite Sorbents-NH₃

Using ENG-TSA as the matrix, various chemical sorbent composite materials are tested and studied. The circulating sorption capacity and required sorbent weight of different composite sorbents are shown in Table 5.13. The density of the sorbent is 400 kg/m³, the ratio of solid sorbent to ENG-TSA is 4:1, the sorption temperature is 20 °C and the desorption pressure is 4 bar [101]. For the data in Table 5.13, the equilibrium temperature, equilibrium pressure and sorption capacity are tested, and the Clapeyron curve of the experimental data is obtained. The sorbent mass of the Type A car is calculated according by Eq. 5.32, and the required ammonia mass is 30–50 g per 100 km. Using the same method, the sorption capacity of the Type B car can also be obtained.

The results show that, except for the $BaCl_2$ + ENG-TSA composite sorbent, all composite sorbents have similar large ammonia cycle sorption capacity. According to Table 5.12, different sorbents adapt to different temperature ranges. In order to find the best composite sorbent in different regions, the weight and volume of the sorbent, the local environmental temperature and the operating cost of the entire system need to be considered.

As shown in Table 5.12, the minimum weight of chemical sorbent per 100 km for Type A trucks is 31.5 g of NH_4Cl per 100 km. The maximum weight is 76.6 g of $BaCl_2$ per 100 km. The same 10,815.8 km mileage requires 6.6 kg NH_4Cl-NH_3 working pair (3.4 kg NH_4Cl and 3.2 kg NH_3) and 13.7 kg $BaCl_2$-NH_3 working pair (8.3 kg $BaCl_2$ with 5.4 kg NH_3), respectively 44.5 and 73.3% lower than AdBlue.

Generally speaking, in the hot summer, the temperature of the truck underframe can reach 50 °C. As the temperature of the sorption bed increases, the pressure of the ASS device increases. At 50 °C, the pressure of LTS can reach 10 bar, which is 250% higher than 4 bar. The higher the LTS pressure, the thicker the stainless steel wall of the ASS device. Ultimately, it will increase the loading capacity of the

Table 5.13 Cycle sorption capacity and the sorbent mass required by different types of trucks per 100 km [97]	Composite sorbents	Cycle sorption quantity (kg/kg)	Sorbent of type-A (g)	Sorbent of type-B (g)
	NH_4Cl + ENG-TSA [101]	0.636 (4 bar)	47.2–78.7	236.0–393.3
	NaBr + ENG-TSA	0.704 (4 bar)	42.6–71.0	213.1–355.2
	$BaCl_2$ + ENG-TSA	0.379 (4 bar)	79.2–132.1	396.2–660.4
	$CaCl_2$(8–2) + ENG-TSA [101]	0.613 (4 bar)	49.0–81.6	244.9–408.1
	$SrCl_2$ + ENG-TSA [101]	0.606 (4 bar)	49.5–82.5	247.6–412.6

truck. Therefore, LTSs have the characteristics of low desorption temperature and light weight of sorbents, which are suitable for cold areas ($\leq$10 °C). MTSs are more suitable for normal weather when the temperature is generally above 10 °C.

Compared with ammonia sorption refrigeration, the diameter of the ASS tank containing the sorbent should be larger due to the requirement for the maximum capacity of ammonia. In addition, it is also proved that the solid composite sorbent with a mass ratio of 4:1 has the best thermal conductivity and permeability, so 80% halide and 20% ENG-TSA are selected for sorbent preparation and sorption/desorption experiments as important prerequisites.

The bulk volume and weight of different composite ammonia sources are compared in Figs. 5.69 and 5.70 for type-A truck and type-B truck. The bulk volume of AdBlue required is:

$$V_{A\_B} = V_{D\_O} \times (3 - 5\%) \tag{5.41}$$

where V_{A_B} and V_{D_O} are the bulk volume of AdBlue and diesel oil separately for 100 km mileage.

The volume of the composite sorbent can be calculated from its density and the relevant sorbent mass mentioned in Table 5.13. Since the emission of NO$_x$ pollutants

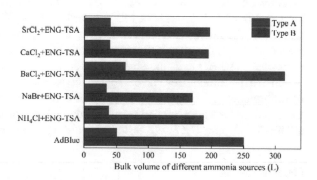

Fig. 5.69 Bulk volume comparisons of different ammonia sources of two types diesel vehicles [97]

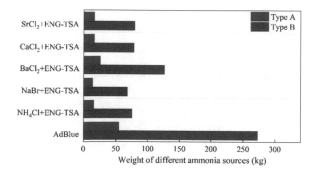

Fig. 5.70 Weight comparisons of different ammonia sources of two types diesel vehicles [97]

is very small, the desorption rate of NH$_3$ is very small during the long desorption stage. For example, for a Type A truck, the NH$_3$ required per 10,000 km is only 3–5 kg. When the truck runs at 60 km/h, the average desorption rate is only 0.018–0.03 kg/h. Therefore, the influence of the desorption rate that meets the NO$_x$ release rate requirement is negligible. Taking a bucket of AdBlue (about 50 L) as a benchmark, the data shown in Fig. 5.69 and Fig. 5.70 are obtained and compared.

As shown in Fig. 5.69 and Fig. 5.70, the volume and weight results of NH$_4$Cl + ENG-TSA, NaBr + ENG-TSA, CaCl$_2$ + ENG-TSA and SrCl$_2$ + ENG-TSA are similar, and the difference is small. The volume of AdBlue (Fig. 5.69) is similar to that of the composite chemical sorbent, but due to the higher density of the AdBlue solution, the weight of AdBlue (Fig. 5.70) is much larger than that of the composite chemical sorbent. In addition, an additional urea pump is always required in the urea SCR system. Therefore, composite halide is the best choice for trucks. On the other hand, the pyrolysis temperature of urea is 160 °C. When the exhaust gas temperature is lower than 160 °C, the urea solution cannot be completely pyrolysed, causing NO$_x$ pollutants to escape, and the average de-NO$_x$ efficiency is reduced.

Figure 5.71 theoretically analyzes the comparison of the NO$_x$ conversion rate of composite sorbent SCR and urea SCR in the temperature range of 50–250 °C. In Fig. 5.71, the NO$_x$ conversion rate is considered to be the best when NH$_3$/NO$_x$ = 0.8–1 [102], and the de-NO$_x$ performance of different catalysts for the NH$_3$-SCR reaction can be seen in reference [103]. In the urea SCR reaction, the pyrolysis temperature of AdBlue is about 160 °C, starting at 160 °C to generate NH$_3$ and react with NO$_x$. Assuming that AdBlue is completely pyrolyzed when the temperature reaches 250 °C, the urea SCR and composite sorbent SCR have the same de-NO$_x$ performance when the temperature is higher than 250 °C.

Table 5.14 summarizes and compares the permeability and thermal conductivity of different solid CaCl$_2$ sorbents. Under the condition of a density of 400 kg/m^3, the sorption/desorption properties of two composite sorbents (80% CaCl$_2$ and 20% additive) and pure CaCl$_2$ are tested. According to the experimental data listed in

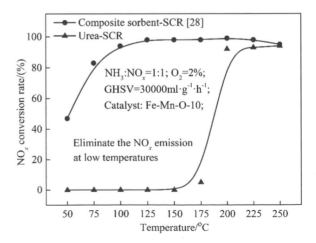

Fig. 5.71 Comparison of NO$_x$ conversion between composite sorbent-SCR and urea-SCR [97]

Table 5.14 Performance comparison between different solid-state CaCl$_2$ sorption materials of 6 mol/mol cycle sorption quantity [97]

Sorption material (Solid State)	Density (kg/m^3)	Mass ratio of salt and additive	Permeability (m^2)	Thermal conductivity (W/(m K))
CaCl$_2$ [49]	400	1:0	5.78×10^{-16}	0.81
CaCl$_2$ + ENG [94]	400	4:1	8.34×10^{-12}	0.89
CaCl$_2$ + ENG-TSA [49]	400	4:1	4.17×10^{-11}	15.10

Table 5.14, the permeability and thermal conductivity of CaCl$_2$ + ENG-TSA are both the highest, indicating that the addition of ENG-TSA helps the improvement of heat and mass transfer. This effect of the additional matrix is due to the microporous structure and ordered layered structure of ENG-TSA [94].

In addition, Fig. 5.71 shows the NO$_x$ conversion effect between the composite sorbent SCR (using CaCl$_2$ + ENG-TSA as the composite sorbent) and urea SCR at different temperatures at a gas hourly space velocity (GHSV) of 30,000 ml/(g h). The NO$_x$ conversion effect is compared in Fig. 5.71. In the case of NH$_3$:NO$_x$ = 1:1 and oxygen content of 2%, the NO$_x$ conversion rate of the composite sorbent-SCR system at 50 °C is about 45%, while the NO$_x$ conversion rate of the urea-SCR system under the same conditions is zero. The area between the composite sorbent SCR line and the urea SCR line in Fig. 5.71 represents the de-NO$_x$ advantage of the composite sorbent SCR system at low temperatures. This area represents the increase in the average NO$_x$ conversion rate of the composite sorbent SCR system. The ammonia content in the ammonia source, the degree of pyrolysis of ammonia-containing materials and the catalytic conversion effect are the main factors that affect the low-temperature NO$_x$ conversion effect of the NH$_3$-SCR system.

The different components of composite sorbent SCR and urea SCR are shown in Fig. 5.72. For the composite sorbent SCR, the ASS unit consists of a stainless steel ASS tank and the solid sorbent saturated with NH$_3$. For urea SCR, the corresponding parts of the ASS device include AdBlue tank, AdBlue solution and AdBlue pump. Regarding the performance of these two systems, the requirement of urea for heating

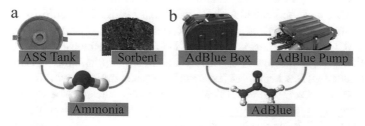

Fig. 5.72 Different components between **a** composite sorbent-SCR and **b** urea-SCR [97]

temperature (above 160 °C) is much higher than that of composite sorbent NH_3 (50–100 sC). For 32.5% urea solution, the theoretical ammonia content of urea is about 18.42%. Taking the performance of the two systems into account, for the composite sorbent SCR cycle, the working medium consumed is ammonia (100% ammonia content), and its price is lower than that of urea solution. In summary, the NO_x conversion performance of the composite sorbent SCR realizes the low-temperature de-NO_x effect and greatly improves the utilization rate for the heat of exhaust gas.

5.4 Solid Sorption Cycle for Heat Transfer

Heat transfer involves various energy utilization processes. Heat pipe (HP) is an efficient and reliable passive heat exchange device that uses capillary force to achieve a continuous evaporation/condensation cycle with small temperature drop. In the past few decades, due to different requirements for thermal control and cooling systems, heat pipes have been significantly improved in achieving higher heat flux [104, 105].

The heat transfer capability of HP is limited by several heat transfer limitations, mainly including viscosity, sonic, capillary, entrainment, and boiling limit [106, 107]. The limitation of the core structure is the key factor. As the core component of the HP, the wick structure has two major functions of heat and mass transfer. First, it generates capillary pressure to transport the working fluid from the condenser section to the evaporator section. Second, it provides a way for liquid distribution. The equivalent thermal conductivity, permeability and maximum capillary pressure are related to the capillary structure. The maximum heat transfer capacity and minimum heat flow resistance of HPs are discriminated by experiments, which are completely related to the capillary structure inside [104].

A coreless structure pipe in which the evaporation section must be vertically located below the condensation section is called a two-phase closed thermosiphon [108]. They are widely used in waste heat recovery systems [109], solar heating systems [110] and thermal management components [111]. For thermosiphons with high heat input, the liquid–vapor (L-V) interface near the exit of the evaporator (the highest vapor flow rate) will become agitated and wavy. This process reduces the speed at which the liquid returns to the evaporator at the expense of increasing the thickness of the liquid film. The mismatch between the rate of liquid returning to the evaporator and the heat input may eventually cause the liquid film drying in the evaporator section. This is called the "onset of flooding" or counter current flow limitation (CCFL) of the thermosiphon [112–114] as the condenser becomes flooded.

Sorption is a general term that includes solid sorption and absorption. Sorption reaction can adopt low heat level (below 150 °C) and optional working medium. In recent years, solid sorption technology has received widespread attention in large-scale or long-term thermal energy storage due to its high reaction heat and energy density [115, 116]. It can use environmentally friendly refrigerants with zero global warming potential and use solar energy or waste heat as the main energy source

[117]. However, solid sorption processes have never been proposed for continuous heat transfer, mainly because they are intermittent.

Most previous studies for heat pipe did not concentrate on the solid sorption principle for substituting the wick and working fluid in conventional HP. Vasiliev et al. earlier proposed sorption HP in previous publications [118, 119]. They combined the conventional HP with the desorption and sorption phases of solid sorbents. In his work, the desorption phase can be completed with a conventional HP's condensing phase, and the sorption phase can be used as the storage phase of liquid fluid of conventional HP. The desorption and refrigeration process of the sorption system adopts conventional HP. Because the desorption and sorption phases need to be switched, the working phase is intermittent.

Combining the desorption between solid and gas, the sorption between liquid and gas, and condensation process, a new concept of solid sorption heat pipe (SSHP) is proposed. It can realize continuous heat transfer, and overcome the shortcomings of conventional HP and thermosiphon. Experiments are conducted under different operating conditions to investigate the overall performance of SSHP, and verify that the test device can be effectively and continuously utilized for transfering thermal energy.

5.4.1 Fundamentals of SSHP

To clearly illustrate the advantages of SSHP, Fig. 5.73 shows two types of SSHP (vertical and horizontal). The SSHP in Fig. 5.73a transfers thermal energy from the heat source to the radiator vertically, which is similar to a conventional thermosiphon.

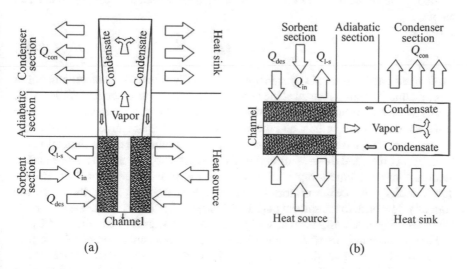

Fig. 5.73 Structure and operation of SSHP. **a** Vertical; **b** horizontal [120]

The SSHP in Fig. 5.73b transfers heat horizontally. For this type of conventional HP, a core structure is essentially required. In Fig. 5.73a, b, the solid sorbent is filled in the sorbent part, that is, the conventional HP evaporator part, and sorbate is sorbed in the solid sorbent. In other words, use solid sorbents and sorbate to replace conventional HP working fluids. The heat transfer mechanism of SSHP can be divided into two working stages:

(1) Heating and desorption stage. At this stage, the sorbent is partially heated, thereby providing desorption heat (Q_{des} in Fig. 5.73). Then, the sorbate is desorbed and the steam flows to the condenser section through the steam channel.
(2) Condensation and liquid reflux stage. At this stage, the vapor in the condenser part is cooled by the cooling fluid, so the film condensation starts from the inner wall and the vapor sorbate condenses into a liquid working fluid. The latent heat of evaporation is released to the heat sink (Q_{con} in Fig. 5.73). After that, the condensate flows back to the sorbent part along the inner wall under the action of gravity (Fig. 5.73a) or sorption effect (Fig. 5.73b) and is sorbed there by the solid sorbent.

These two phases ensure a continuous heat transfer process. Three types of heat exist in these two working phases as Fig. 5.73a, b show: desorption heat (Q_{des}) between solid sorbent and vapor sorbate, reaction heat of the solid sorbent and liquid sorbate (Q_{l-s}), and condensation heat (Q_{con}). The reaction between the solid sorbent and liquid sorbate will release heat and the bonding effect of Q_{des} and Q_{l-s} equals to Q_{in} (heat input), i.e.:

$$Q_{in} = Q_{des} - Q_{l-s} \tag{5.42}$$

Taking the whole SSHP as one system, the heat input should be equal to the heat output, i.e.:

$$Q_{in} = Q_{con} \tag{5.43}$$

As mentioned earlier, the thermosiphon is usually filled with a small amount of working fluid, which is located at the bottom of the evaporation section, while CCFL occurs at the end of the evaporator, which is the liquid film inlet area and the steam outlet area. The solid sorbent and the sorbate in SSHP are completely filled in the sorbent part. The desorbed steam passes through the mass transfer channel under the heating effect and diffuses around in the adiabatic section and the condensation section. Since the liquid–vapor interface is not easy to agitate and fluctuate, the flow velocity at the center is large and is small near the wall. Thus, the liquid can normally flow back to the sorbent part. In this way, SSHP can alleviate the restriction of conventional thermosiphons caused by the countercurrent flow of vapor and liquid.

For the horizontal SSHP shown in Fig. 5.73b, it uses a compound of solid sorbent and sorbate in the evaporation part of HP, which is different from the core structure of conventional HP. Taking the liquid reflux phase into account, the pressure difference between the sorption part and the condenser provided by the sorbent sorption will drive the liquid to flow from the condensation section to the sorption section, rather than the capillary force of the core. In this way, the heat transfer limit of conventional HP caused by the liquid–gas medium and the capillary structure can be avoided. Compared with the conventional HP, the vertically placed SSHP does not have the possibility of downward heat transfer, because the porous material distributed along the entire length of the heat pipe is replaced by the porous sorbent in the single-stage evaporator. Therefore, the vertical SSHP is a thermosiphon filled with sorbent in the evaporator section, which is a limitation compared with conventional HP.

5.4.2 Experimental Setup

The structure of the SSHP machine and the experimental test device are shown in Fig. 5.74. The test device (Fig. 5.74a) mainly includes an SSHP, two flow meters, two constant temperature baths, a pressure transmitter, six platinum resistance thermometers and valves. The prepared composite sorbents (NaBr and ENG-TSA) are filled in the space of the sorbent section, and then compressed into a consolidated block. After that, the sorption section is connected with the adiabatic section and the condensation section. Before the experiment, the SSHP is kept in a vacuum state, and then a certain amount of NH_3 is filled into the SSHP through the filling tube. The sorbent block exchanges heat with the heat transfer fluid (heating fluid) through the external heat transfer jacket, and the sorbent vapor is desorbed from the sorbent block through the vapor channel in the middle of the block. In the condenser section, the sorbate steam transfers latent heat to the heat transfer fluid (cooling fluid) through the external jacket. At the same time, the cooling sorbate liquid returns to the sorbent through the inner wall, and the heat transfer from the sorption section to the condenser section is realized through the process of desorption and condensation. The photo of the test device is shown in Fig. 5.74b.

In order to quantitatively study the performance of the SSHP machine, a certain amount of solid composite sorbent and sorbate are pre-added to the stainless steel tube. Experiments are conducted under different operating conditions to study the influence of various parameters on the overall performance of the SSHP unit. The controlled conditions in the test are the inlet hot water temperature of the heat source and the inlet cold water temperature of the heat sink.

In fact, the two heat transfer processes between the SSHP and the heat source/sink are carried out by two independent water circulation loops. The thermal energy input from the heat source is divided into three parts: sensible heat, heat required for desorption, and reaction heat between liquid and solid sorbent (Eq. 5.44). In the experiment, the heating temperature gradually increases, and the cooling temperature is controlled within a certain range. Therefore, the sensible heat increases with the increase of the heating temperature, and occupies a considerable part of the heat input in the dynamic equilibrium process. Taking the influence of sensible heat on heat transfer into account, the radial heat transfer quantity and radial heat flux density of the sorbent and condenser can be obtained by using Eqs. 5.45–5.48 respectively.

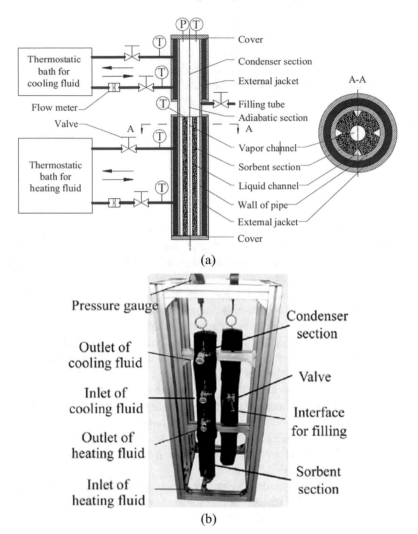

Fig. 5.74 Experimental test system of SSHP. **a** Schematic diagram; **b** photo [120]

$$Q_{in} = Q_{des} + Q_{sens} - Q_{l-s} \tag{5.44}$$

It should be noted that Q_{l-s} cannot be neglected though it is much smaller than the Q_{des}. For example, if the reaction occurs under the conditions of an isothermal process, sensible heat can be ignored. Then according to Eqs. (5.43) and (5.44), the Q_{l-s} is the difference between Q_{des} and Q_{cond}. Taking the theoretical equilibrium data to analyze the results, Q_{des} (reaction heat between the gas and the solid) of NaBr-NH_3 is about 1790 kJ/kg and the condensation heat of NH_3 is 1186 kJ/kg when the condensing temperature is 20 °C, so the Q_{l-s} is 604 kJ/kg, which is around 1/3 of the total desorption heat of Q_{des}.

$$Q_{sor} = Q_{in} = C_p \rho \dot{V} \left(T_{hw,in} - T_{hw,out} \right) + \left(Q_{sens,cs} + Q_{sens,s} + Q_{sens,ss} \right)$$
$$= C_p \rho \dot{V} \left(T_{hw,in} - T_{hw,out} \right) + \left(C_{p,cs} m_{cs} + C_{p,s} m_s + C_{p,ss} m_{ss} \right)_{sor} \frac{dT_{sor}}{dt} \quad (5.45)$$

$$Q_{con} = C_p \rho \dot{V} \left(T_{cw,out} - T_{cw,in} \right) \quad (5.46)$$

$$q_{rad,sor} = Q_{des} / A_{sor} \quad (5.47)$$

$$q_{rad,con} = Q_{con} / A_{con} \quad (5.48)$$

where Q_{sens} is the total sensible heat (kW), C_p is the specific heat of water (kJ/(kg K)), ρ is the density of water (kg/m^3), $\dot{V}$ is the volumetric flow rate of water (m^3/s), m is the mass (kg), T_{in} and T_{out} are the inlet temperature of water and the outlet temperature of water respectively (°C), t is time (s), q is the heat flux (kW/m^2) and A is the heat transfer area (m^2). The subscripts of hw, cw, cs, s, ss, sor, con and rad are representative for hot water, cold water, composite sorbent, sorbate, stainless steel, sorbent section, condenser section and radial respectively.

The heat transfer medium inside SSHP is ammonia sorbate. In the sorption part, heat is transferred through the desorption process of ammoniate halide. In the condenser section, heat is transferred through the condensation process of ammonia vapor. In order to simplify the analysis, the axial heat flux based on the condensation heat of SSHP can be calculated by Eq. 5.49 as follows:

$$q_{axi} = Q_{con} / A_c \quad (5.49)$$

The heat transfer capacity per unit molar ammonia and thermal resistance can be obtained by Eqs. 5.50 and 5.51 as follows:

$$Q_{con} = Q_{con} / n \quad (5.50)$$

$$R = \Delta T / Q_{in} = \Delta T / Q_{con} \quad (5.51)$$

where A_c is the cross-section area of SSHP (m^2), Q_{con} is the heat transfer quantity per unit molar of ammonia (W/mol), n is the molar amount of ammonia (mol), R is the thermal resistance (K/W), ΔT is the temperature difference between the sorbent section and condenser section (K) and the subscript of axi is representative for axial.

5.4.3 Thermal Performance of SSHP

5.4.3.1 Vertical SSHP

Under different experimental conditions, the total heat transfer can be used to characterize the thermal performance of the SSHP unit. The cooling temperatures of 20, 25 and 30 °C are studied, and the results are shown in Fig. 5.75. The heat transfer quality of the sorption section increases with the increase of heating temperature. When the heating temperature is lower than 60 °C, the heat transfer increases slowly. While in the range of 60–80 °C, due to the occurrence of the desorption process, the heat transfer rate increases significantly, meanwhile the sorbate vapor and condensate flowrate increases, which transfers large amount of heat. In addition, it can be seen that under the same conditions, the overall heat transfer performance of the cooling temperature of 25 °C is close to that of 30 °C. But when the cooling temperature is 20 °C, the heat transfer performance is much better. When the heating temperature is 80 °C and the cooling temperature is 20 °C, the heat transfer capacity of SSHP reaches the maximum value of 1.50 kW.

The relationship between the radial heat flux and heating temperature is shown in Fig. 5.76. From 60 to 80 °C, the heat flux of the sorbent section and the condenser section both increase significantly, and the variation of q_{con} is approximately linear. As shown in Fig. 5.76a, when the heating temperature is 90 °C and the cooling temperature is 20 °C, the values of q_{con} and q_{sor} are 22.1 kW/m^2 and 18.5 kW/m^2, respectively. When the cooling temperature is 25 °C, the highest heat transfer fluxes of q_{con} and q_{sor} are 21.8 kW/m^2 and 14.0 kW/m^2, respectively, which are 1.4 and 24.3% less than when the cooling temperature is 20 °C. The results show that at the cooling temperature of 25 °C, the value of q_{con} is relatively close to 20 °C, while the q_{sor} at 25 °C is smaller than the q_{sor} at 20 °C. Figure 5.76a, b both show that, in the range of 60–80 °C, the difference between the condensation section and the

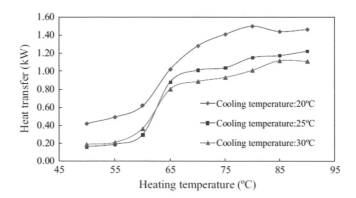

Fig. 5.75 Heat transfer in sorbent section of vertical SSHP versus heating temperature [120]

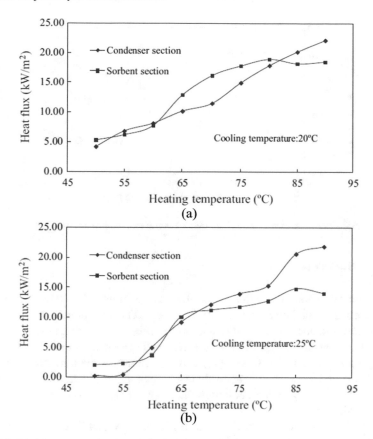

Fig. 5.76 Heat flux in sorbent and condenser section of vertical SSHP. **a** Cooling temperature of 20 °C; **b** cooling temperature of 25 °C [120]

sorbent section is less than 29.1%. This result is consistent with the energy balance calculation of the system.

In addition to the thermal performance between the SSHP unit and the heat source/heat sink, it is also necessary to investigate the internal heat transfer of the SSHP characterized by axial heat flow and internal temperature difference. Since the heat input of the sorbent part is almost equal to the heat output of the condenser part, in order to simplify the analysis, Fig. 5.77 plots the axial heat flux results based on the condenser part. The results show that when the cooling temperature is 30 °C, the entire heat flux density is less than that of 20 and 25 °C, while the heat flux value at 25 °C is close to 20 °C. At a heating temperature of 90 °C, the maximum axial heat flux inside the SSHP is about 800 kW/m^2.

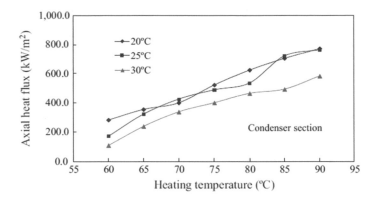

Fig. 5.77 Axial heat flux of vertical SSHP versus heating temperature [120]

5.4.3.2 Horizontal SSHP

Taking typical working conditions (cooling temperature 20 °C) as an example, the performance of horizontal SSHP is tested. The calculation results of heat transfer performance and internal temperature difference are given in the Figs. 5.78 and 5.79. As shown in Fig. 5.78, when the heating temperature is 80 °C and the cooling temperature is 20 °C, the q_{con} and q_{sor} values of the horizontal SSHP are 12.2 kW/m^2 and 12.4 kW/m^2, respectively, which are 31.5 and 34.4% lower than the vertical SSHP (Fig. 5.76a), although the trends in the two graphs (Figs. 5.78 and 5.76a) are similar.

When the heating temperature reaches 60 °C or above, as shown in Fig. 5.79, the axial heat flux increases slowly with the increase of temperature. It can be inferred that the operating speed of the desorption and condensing working cycle in horizontal SSHP is low, and the heat exchange capacity is quickly approaching the peak value.

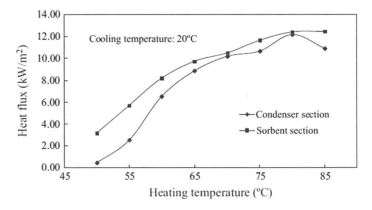

Fig. 5.78 Heat flux in sorbent section and condenser section of horizontal SSHP [120]

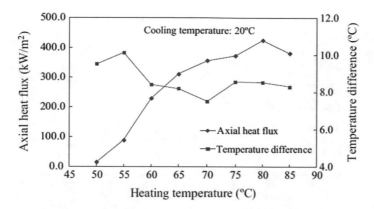

Fig. 5.79 Axial heat flux and inner temperature difference of horizontal SSHP [120]

The maximum axial heat flux of the horizontal SSHP is about 425 kW/m². Compared with the thermal performance of the vertical SSHP in Fig. 5.77, when the heating temperature is higher than 75 °C, the difference between the vertical SSHP and the horizontal SSHP becomes larger. It can be concluded that the vertical SSHP has better overall transmission than the horizontal SSHP. According to the experimental results, the NaBr-NH₃ working pair is suitable for low-grade heat transfer above 60 °C. Basic test results show that the non-isothermal heat transfer performance of SSHP is significantly different from conventional HP.

5.4.3.3   Heat Transfer Performance of SSHP with Different Amounts of Sorbate

The heat transfer capacity of SSHP depends on basic parameters such as pressure, temperature and flow rate. Figure 5.80 shows the heat transfer results of 3 mol of

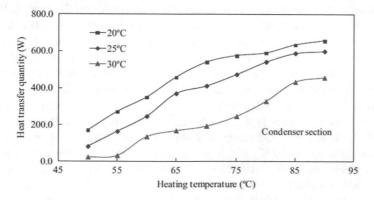

Fig. 5.80 Heat transfer capacity in condenser section of SSHP with 3 mol sorbates [121]

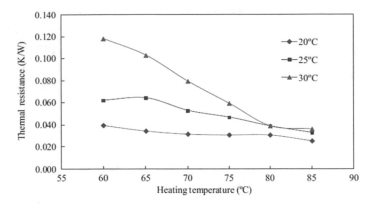

Fig. 5.81   The overall thermal resistance of SSHP with 3 mol sorbates [121]

sorbate at different cooling temperatures. For a cooling temperature of 20 °C, the heat transfer amount increases approximately linear in the entire heating temperature range (50–90 °C), which indicates that the condensation rate of the desorbed sorbate is relatively large under the corresponding condensation pressure. However, at a cooling temperature of 30 °C, the condensation process is relatively more difficult than that of 20 °C. Therefore, the smaller the condensation rate, the smaller the heat exchange.

Figure 5.81 shows the effect of heating and cooling temperature on the thermal resistance between the condenser and the sorbent section. According to the previous analysis in Fig. 5.81, the heat transfer capacity (the three curves in Fig. 5.80) increases significantly within the heating temperature range of 60–85 °C. Therefore, the thermal resistance is calculated, and the calculation result for this temperature range is shown in Fig. 5.81. The experimental results in Fig. 5.80 show that the heat transfer increases with the heating temperature, and the thermal resistance in Fig. 5.81 decreases with the increase of heating temperature. Therefore, according to Eq. 5.51, the increase in temperature difference with heating temperature is expected to be smaller than the increase in heat transfer capacity. When the cooling temperature is 30 °C, the thermal resistance is significantly higher than 20 and 25 °C. This phenomenon shows that at a certain heating temperature and 30 °C cooling temperature, although the temperature difference between the sorption section and the condensation section is less than those of 20 and 25 °C, the heat transfer is much smaller. This leads to an increase in the ratio of temperature difference and heat transfer, that is, at a cooling temperature of 30 °C, the overall thermal resistance of SSHP is higher than other types. In addition, compared with the three sets of data (Fig. 5.81), the size and fluctuation of thermal resistance are relatively small at the cooling temperature of 20 °C, which proves that SSHP can operate well in this case and has a stable non-isothermal heat transfer performance. Another point in Fig. 5.81 is that at higher heating temperatures, the three curves are close to a relatively low thermal resistance value, which is a useful guide for practical applications and operations. The thermal resistance in the experiment ranges from 0.025 to 0.118 K/W, and

the maximum value appears when the cooling temperature is 30 °C and the heating
temperature is 60 °C.

The working condition of cooling temperature of 20 °C is selected to further investigate the characteristics of SSHP with different amounts of ammonia. Figure 5.82
plots the relationship between the radial heat flux of the sorbent and the heating
temperature. From 60 to 90 °C, the heat flux density increases with the increase of
heating temperature, and it changes approximately linearly. In addition, the slope of
the 5 mol curve is the largest. When the heating temperature is 90 °C and the cooling
temperature is 20 °C, the maximum heat flux density is 14.84 kW/m^2.

In order to better understand the overall performance of SSHP, the results of
heat transfer per mole of ammonia and axial heat flux are calculated numerically in
Figs. 5.83 and 5.84. As shown in Fig. 5.83, the heat transfer capacity per unit molar
ammonia for NaBr-1NH$_3$ is larger than that of 3 and 5 mol, that is, the increase in
heat transfer is less than the increase in the filling amount of ammonia. When the

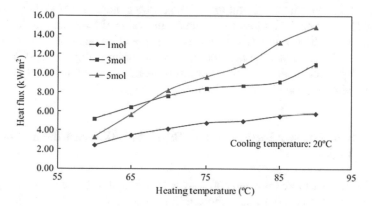

Fig. 5.82 Heat flux in sorbent section with cooling temperature of 20 °C [121]

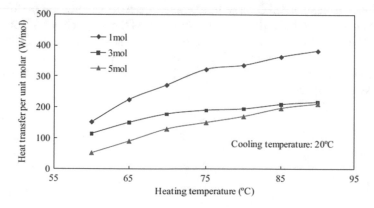

Fig. 5.83 Heat transfer per unit molar of ammonia with cooling temperature of 20 °C [121]

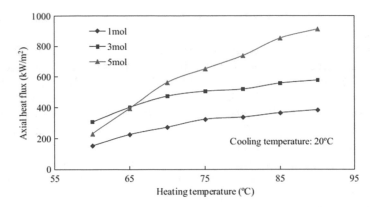

Fig. 5.84 Axial heat flux of SSHP with cooling temperature of 20 °C [121]

cooling temperature is 20 °C and the heating temperature is 90 °C, the maximum value of 3 mol ammonia is close to the maximum value of 5 mol ammonia. It can be seen from Fig. 5.84 that the axial heat flux of 1 mol ammonia is much smaller than that of 3 and 5 mol. Under the same conditions of cooling temperature of 20 °C and heating temperature of 90 °C, the maximum value of axial heat flux is 913.3 kW/m², while the minimum value is 384.9 kW/m².

5.4.3.4   Heat Transfer Performance of SSHP with Different Inclination Angles

Figure 5.85 shows more experimental results of different SSHP arrangements (inclination = 90°, 45° and 0°) under the cooling temperature of 20 and 35 °C. Take SSHP containing 3 mol of sorbent as an example. In addition to the change trend of heat transfer capacity consistent with previous research results, it also shows that the inclination angle of SSHP has a significant impact on heat transfer performance. From the curves of (a) and (b), it can be seen that as the angle changes from 90° to 45°, the axial heat flux decreases more than that of the angle changes from 45° to 0°, especially at higher heating temperatures. When the cooling temperature is 20 °C, the maximum axial heat flux of 0°, 45° and 90° is 354.9 kW/m², 403.0 kW/m² and 559.7 kW/m², respectively.

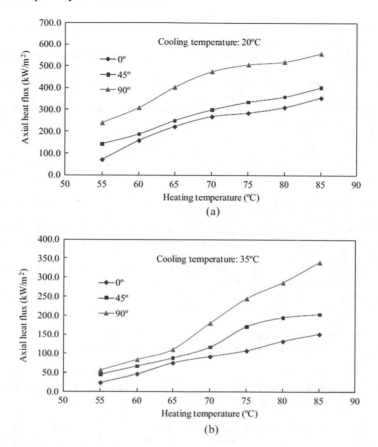

Fig. 5.85 Heat transfer performance with different inclination angles **a** 20 °C and **b** 35 °C [121]

Compared with ammonia heat pipes studied in recent years [122–127], SSHP has an advantage in overall thermal performance when the heat transfer medium, geometry and operating conditions are similar.

References

1. Restuccia G, Freni A, Vasta S, Yu A (2004) Selective water sorbent for solid sorption chiller: experimental results and modelling. Int J Refrig 27:284–293
2. Saha BB, Koyama S, Ng KC, Hamamoto Y, Akisawa A, Kashiwagi T (2006) Study on a dual-mode, multi-stage, multi-bed regenerative sorption chiller. Renew Energy 31:2076–2090
3. Wang DC, Xia ZZ, Wu JY, Wang RZ, Zhai H, Dou WD (2005) Study of a novel silica gel-water sorption chiller. Part I. Design and performance prediction. Int J Refrig 28:1073–1083
4. Saha BB, Kashiwagi T (1997) Experimental investigation of an advanced sorption refrigeration cycle. Ashrae Trans 103:50–58

5. Saha BB, Akisawa A, Kashiwagi T (2014) Solar/waste heat driven two-stage sorption chiller: the prototype. Renew Energy 23:93–101
6. Erhard A, Hahne E (1997) Test and simulation of a solar-powered absorption cooling machine. Sol Energy 59:155–162
7. Erhard A, Spindler K, Hahne E (1998) Test and simulation of a solar powered solid sorption cooling machine. Int J Refrig 21:133–141
8. Pons M, Guilleminot JJ (1986) Design of an experimental solar-powered, solid-sorption ice maker. J Sol Energy Eng 108:332–337
9. Wang LW, Wang RZ, Lu ZS, Xu YX, Wu JY (2006) Split heat pipe type compound sorption ice making test unit for fishing boats. Int J Refrig 29:456–468
10. Meunier F (1986) Theoretical performances of solid adsorbent cascading cycles using the zeolite-water and active carbon-methanol pairs: four case studies. J Heat Recover Syst 6:491–498
11. Pierrès NL, Mazet N, Stitou D (2007) Experimental results of a solar powered cooling system at low temperature. Int J Refrig 30:1050–1058
12. El-Sharkawy II, Uddin K, Miyazaki T, Saha BB, Koyama S, Miyawaki J, Yoon S-H (2014) Sorption of ethanol onto parent and surface treated activated carbon powders. Int J Heat Mass Transf 73:445–455
13. El-Sharkawy II, AbdelMeguid H, Saha BB (2013) Towards an optimal performance of sorption chillers: reallocation of sorption/desorption cycle times. Int J Heat Mass Transf 63:171–182
14. Walmsley TG, Walmsley MRW, Atkins MJ, Neale JR (2014) Integration of industrial solar and gaseous waste heat into heat recovery loops using constant and variable temperature storage. Energy 75:53–67
15. Ali SM, Chakraborty A (2015) Thermodynamic modelling and performance study of an engine waste heat driven sorption cooling for automotive air-conditioning. Appl Therm Eng 90:54–63
16. Sharafian A, Mehr SMN, Thimmaiah PC, Huttema W, Bahrami M (2016) Effects of adsorbent mass and number of sorbent beds on the performance of a waste heat-driven sorption cooling system for vehicle air conditioning applications. Energy 112:481–493
17. Zhong YF, Fang TG, Wert KL (2011) An sorption air conditioning system to integrate with the recent development of emission control for heavy-duty vehicles. Energy 36:4125–4135
18. Mills A, Farid M, Selman JR, Al-Hallaj S (2006) Thermal conductivity enhancement of phase change materials using a graphite matrix. Appl Therm Eng 26:1652–1661
19. Wang LW, Tamainot-Telto Z, Metcalf SJ, Critoph RE, Wang RZ (2010) Anisotropic thermal conductivity and permeability of consolidated expanded natural graphite. Appl Therm Eng 30:1805–1811
20. Fayazmanesh K, Salari S, Bahrami M (2017) Effective thermal conductivity modeling of consolidated sorption composites containing graphite flakes. Int J Heat Mass Transf 115:73–79
21. Tanashev YY, Krainov AV, Aristov YI (2013) Thermal conductivity of composite sorbents "salt in porous matrix" for heat storage and transformation. Appl Therm Eng 61:401–407
22. Wang LW, Metcalf SJ, Critoph RE, Thorpe R, Tamainot-Telto Z (2011) Thermal conductivity and permeability of consolidated expanded natural graphite treated with sulphuric acid. Carbon 49:4812–4819
23. Gao J, Wang LW, An GL, Liu JY, Xu SZ (2018) Performance analysis of multi-salt sorbents without sorption hysteresis for low-grade heat recovery. Renew Energy 118:718–726
24. Gao J, Wang LW, Gao P, An GL, Lu HT (2018) Design and analysis of a gas heating/cooling sorption refrigeration system with multi-salt solid sorbent of $CaCl_2$ and $MnCl_2$. Int J Heat Mass Transf 126:39–47
25. Gao J, Wang LW, Gao P, An GL, Wang ZX, Xu SZ, Wang RZ (2019) Performance investigation of a freezing system with novel multi-salt sorbent for refrigerated truck. Int J Refrig 98:129–138
26. Wang LW, Wang RZ, Wu JY, Wang K, Wang SG (2004) Sorption ice makers for fishing boats driven by the exhaust heat from diesel engine: choice of sorption pair. Energy Convers Manage 45:2043–2057

27. Neveu P, Castaing J (1993) Solid-gas chemical heat pumps: field of application and performance of the internal heat of reaction recovery process. Heat Recovery Syst CHP 13:233–251
28. Yuanyang HU, Wang L, Lu XU, Wang R, Kiplagat J, Wang J (2011) A two-stage deep freezing chemisorption cycle driven by low-temperature heat source. Front Energy 5:263–269
29. Wang J, Wang LW, Luo WL, Wang RZ (2013) Experimental study of a two-stage adsorption freezing machine driven by low temperature heat source. Int J Refrig 36:1029–1036
30. Chatzidakis SK, Chatzidakis KS (2004) Refrigerated transport and environment. Int J Energy Res 28:887–897
31. Wang RZ, Oliveira RG (2006) Sorption refrigeration-An efficient way to make good use of waste heat and solar energy. Prog Energy Combust Sci 32:424–458
32. Fernandes MS, Brites GJVN, Costa JJ, Gaspar AR, Costa VAF (2014) Review and future trends of solar sorption refrigeration systems. Renew Sustain Energy Rev 39:102–123
33. Hamdy M, Askalany AA, Harby K, Kora N (2015) An overview on sorption cooling systems powered by waste heat from internal combustion engine. Renew Sustain Energy Rev 51:1223–1234
34. Zhang LZ (2000) Design and testing of an automobile waste heat sorption cooling system. Appl Therm Eng 20:103–114
35. Jiangzhou S, Wang RZ, Lu YZ, Xu YX, Wu JY (2002) Experimental investigations on sorption air-conditioner used in internal-combustion locomotive driver-cabin. Appl Therm Eng 22:1153–1162
36. Lu YZ, Wang RZ, Jianzhou S, Xu YX, Wu JY (2004) Practical experiments on an sorption air conditioner powered by exhausted heat from a diesel locomotive. Appl Therm Eng 24:1051–1059
37. Abdullah MO, Tan IAW, Lim LS (2011) Automobile sorption air-conditioning system using oil palm biomass-based activated carbon: a review. Renew Sustain Energy Rev 15:2061–2072
38. Zhang HG, Wang EH, Fan BY (2013) Heat transfer analysis of a finned-tube evaporator for engine exhaust heat recovery. Energy Convers Manag 65:438–447
39. Louajari M, Mimet A, Ouammi A (2011) Study of the effect of finned tube sorbent on the performance of solar driven sorption cooling machine using activated carbon-ammonia pair. Appl Energy 88:690–698
40. Wang RZ, Xia ZZ, Wang LW, Lu ZS, Li SL, Li TX et al (2010) Heat transfer design in sorption refrigeration systems for efficient use of low grade thermal energy. In: 2010 14th international heat transfer conference, pp 575–589
41. Wang D, Zhang J, Yang Q, Li N, Sumathy K (2014) Study of sorption characteristics in silica gel-water sorption refrigeration. Appl Energy 113:734–741
42. Gordeeva L, Frazzica A, Sapienza A, Aristov Y, Freni A (2014) Sorption cooling utilizing the "LiBr/silica - ethanol" working pair: dynamic optimization of the sorbent/heat exchanger unit. Energy 75:390–399
43. Yuan-Yang HU (2011) Performance study of two-stage sorption refrigeration cycle based on $CaCl_2/BaCl_2$-NH_3 working pair. J Eng Thermophys 32:1087–1090
44. Jian W, Yuan-Yang HU, Wang LW, Wang RZ (2011) Experiments on sorption performance and simulation on freezing system for $CaCl_2$-$BaCl_2$-NH_3 two-stage sorption. J Shanghai Jiaotong Univ 45:1389–1394
45. Jiang L, Wang LW, Luo WL, Wang RZ (2015) Experimental study on working pairs for two-stage chemisorption freezing cycle. Renew Energy 74:287–297
46. Gao P, Zhang XF, Wang LW, Wang RZ, Li DP, Liang ZW et al (2016) Study on $MnCl_2/CaCl_2$-NH_3 two-stage solid sorption freezing cycle for refrigerated trucks at low engine load in summer. Energy Convers Manag 109:1–9
47. Gao P, Wang LW, Wang RZ, Zhang XF, Li DP, Liang ZW et al (2016) Experimental investigation of a $MnCl_2/CaCl_2$-NH_3 two-stage solid sorption freezing system for a refrigerated truck. Energy 103:16–26
48. Gao P, Wang LW, Wang RZ, Li DP, Liang ZW (2016) Optimization and performance experiments of a $MnCl_2/CaCl_2$-NH_3 two-stage solid sorption freezing system for a refrigerated truck. Int J Refrig 71:94–107

49. Jiang L, Wang LW, Wang RZ (2014) Investigation on thermal conductive consolidated composite CaCl₂ for sorption refrigeration. Int J Therm Sci 81:68–75
50. Uche J, Martínez-Gracia A, Círez F, Carmona U (2015) Environmental impact of water supply and water use in a Mediterranean water stressed region. J Clean Prod 88:196–204
51. Ziolkowska JR (2014) Is desalination affordable? Regional cost and price analysis. Water Resour Manag 29:1385–1397
52. Fiorenza G, Sharma VK, Braccio G (2003) Techno-economic evaluation of a solar powered water desalination plant. Energy Convers Manag 44:2217–2240
53. Shanmugam GJG, Ravindran S (2004) Review on the uses of appropriate techniques for arid environment. In: International conference on water resources and arid environment
54. Alayli Y, Hadji NE, Leblond J (1987) A new process for the extraction of water from air. Desalination 67:227–229
55. Abualhamayel HI, Gandhidasan P (1997) A method of obtaining fresh water form humid atmosphere. Desalination 113:51–63
56. Bergmair D, Metz SJ, de Lange HC, van Steenhoven AA (2014) System analysis of membrane facilitated water generation from air humidity. Desalination 339:26–33
57. Gandhidasan P, Abualhamayel HI (2005) Modeling and testing of a dew collection system. Desalination 180:47–51
58. Aristov YI (2013) Challenging offers of material science for sorption heat transformation: a review. Appl Therm Eng 50:1610–1618
59. Wang JY, Wang RZ, Wang LW (2016) Water vapor sorption performance of ACF-CaCl₂ and silica gel-CaCl₂ composite adsorbents. Appl Therm Eng 100:893–901
60. Cui Q, Chen H, Tao G, Yao H (2005) Performance study of new adsorbent for solid desiccant cooling. Energy 30:273–279
61. Zheng X, Ge TS, Hu LM, Wang RZ (2015) Development and characterization of mesoporous silicate-LiCl composite desiccants for solid desiccant cooling systems. Ind Eng Chem Res 54:2966–2973
62. Zheng X, Ge TS, Wang RZ (2014) Recent progress on desiccant materials for solid desiccant cooling systems. Energy 74:280–294
63. Yu N, Wang RZ, Lu ZS, Wang LW (2014) Development and characterization of silica gel-LiCl composite sorbents for thermal energy storage. Chem Eng Sci 111:73–84
64. Bui DT, Nida A, Ng Kim C, Chua Kian J (2016) Water vapor permeation and dehumidification performance of poly(vinyl alcohol)/lithium chloride composite membranes. J Membr Sci 498:254–262
65. Aristov YI, Tokarev MM, Cacciola G, Restuccia G (1996) Selective water sorbents for multiple applications, 1. CaCl₂ confined in mesopores of silica gel: sorption properties. React Kinet Catal Lett 59:325–333
66. Aristov YI, Tokarev MM, Restuccia G, Cacciola G (1996) Selective water sorbents for multiple applications, 2. CaCl₂ confined in micropores of silica gel. React Kinet Catal Lett 59:335–342
67. Gordeeva LG, Restuccia G, Cacciola G, Aristov YI (1998) Selective water sorbents for multiple applications, 5. LiBr confined in mesopores of silica gel: sorption properties. React Kinet Catal L 63:81–88
68. Aristov YI, Glaznev IS, Freni A, Restuccia G (2006) Kinetics of water sorption on SWS-1L (calcium chloride confined to mesoporous silica gel): influence of grain size and temperature. Chem Eng Sci 61:1453–1458
69. Aristov YI (2007) Novel materials for adsorptive heat pumping and storage: screening and nanotailoring of sorption properties (review). J Chem Eng Jpn 40:1242–1251
70. Aristov Y, Tokarev MM, Gordeeva LG, Snytnikov VN, Parmon VN (1999) New composite sorbents for solar-driven technology of fresh water production from the atmosphere. Sol Energy 66:165–168
71. Ji JG, Wang RZ, Li LX (2007) New composite adsorbent for solar-driven fresh water production from the atmosphere. Desalination 2:176–182
72. Gad HE, Hamed AM, El-Sharkawy II (2001) Application of a solar desiccant/collector system for water recovery from atmospheric air. Renew Energy 22:541–556

73. Kabeel AE (2007) Water production from air using multi-shelves solar glass pyramid system. Renew Energy 32:157–172
74. Wang JY, Liu JY, Wang RZ, Wang LW (2017) Experimental research of composite solid sorbents for fresh water production driven by solar energy. Appl Therm Eng 121:941–950
75. Hassan HZ, Mohamad AA (2013) Thermodynamic analysis and theoretical study of a continuous operation solar-powered sorption refrigeration system. Energy 61:167–178
76. Liang P, Yuan L, Yang X, Zhou S, Huang X (2013) Coupling ion-exchangers with inexpensive activated carbon fiber electrodes to enhance the performance of capacitive deionization cells for domestic wastewater desalination. Water Res 47:2523–2530
77. Wang JY, Wang RZ, Wang LW, Liu JY (2017) A high efficient semi-open system for fresh water production from atmosphere. Energy 138(nov.1):542–551
78. Riffel DB, Schmidt FP, Belo FA, Leite APF, Cortés FB, Chejne F et al (2011) Sorption of water on Grace Silica Gel 127B at low and high pressure. Sorption 17:977–984
79. Wang JY, Liu JY, Wang RZ, Wang LW (2017) Experimental investigation on two solar-driven sorption based devices to extract fresh water from atmosphere. Appl Therm Eng 127:1608–1616
80. Wang JY, Wang RZ, Tu YD, Wang LW (2018) Universal scalable sorption-based atmosphere water harvesting. Energy 165:387–395
81. Li H, Dai YJ, Li Y, La D, Wang RZ (2011) Experimental investigation on a one-rotor two-stage desiccant cooling/heating system driven by solar air collectors. Appl Therm Eng 31:3677–3683
82. Rosenzweig C, Iglesias A, Yang XB, Epstein PR, Chivian E (2001) Climate change and extreme weather events; implications for food production, plant diseases, and pests. Glob Change Hum Health 2(2):90–104
83. He H, Weng D, Zi XY (2007) Diesel emission control technologies: a review. Environ Sci 28(6):1169–1177
84. Cottle JE (1959) Ammonia storage and recovery system, US
85. Walker J, Speronello BK (1992) Development of an ammonia/SCR NOx reduction system for heavy duty natural gas engine. Int Off Highw Powerpl Congr Expo 931:171–181
86. Chen HJ, Lin MH (1999) Modeling a boiling-liquid, expanding-vapor explosion phenomenon with application to relief device design for liquefied ammonia storage. Ind Eng Chem Res 38:479–487
87. Kaboord WS, Becher DM, Begale FJ, Crane RF, Kuznicki SM (2007) Sorption based ammonia storage and regeneration system, US
88. Chun YL, Aika K (2004) Effect of the Cl/Br molar ratio of a $CaCl_2 - CaBr_2$ mixture used as an ammonia storage material. Ind Eng Chem Res 43:6994–7000
89. Johannessen T, Schmidt H, Svagin J, Johansen J, Oechsle J, Bradley R (2008) Ammonia storage and delivery systems for automotive NOx aftertreatment. SAE World Congr Exhib 2154:1027–1034
90. Veselovskaya JV, Tokarev MM (2011) Novel ammonia sorbents "porous matrix modified by active salt" for adsorptive heat transformation: 4. Dynamics of quasi-isobaric ammonia sorption and desorption on BaCl/vermiculite. Appl Therm Eng 31:566–572
91. Fulks G, Fisher GB, Rahmoeller K, Wu MC, D'Herde E, Tan J (2009) A review of solid materials as alternative ammonia sources for lean NOx reduction with SCR. SAE technical papers
92. Wang LW, Wang RZ, Oliveira RG (2009) A review on sorption working pairs for refrigeration. Renew Sustain Energy Rev 13:518–534
93. Jin ZQ, Wang LW, Jiang L, Wang RZ (2013) Experiment on the thermal conductivity and permeability of physical and chemical compound adsorbents for sorption process. Heat Mass Transf 49:1117–1124
94. Jiang L, Wang LW, Jin ZQ, Wang RZ, Dai YJ (2013) Effective thermal conductivity and permeability of compact compound ammoniated salts in the sorption/desorption process. Int J Therm Sci 71:103–110

95. Zheng X, Ge TS, Wang RZ, Hu LM (2014) Performance study of composite silica gels with different pore sizes and different impregnating hygroscopic salts. Chem Eng Sci 120:1–9

96. Zheng X, Ge TS, Jiang Y, Wang RZ (2014) Experimental study on silica gel-LiCl composite desiccants for desiccant coated heat exchanger. Int J Refrig 51:24–32

97. Wang ZX, Wang LW, Gao P, Yu Y, Wang RZ (2018) Analysis of composite sorbents for ammonia storage to eliminate NOx emission at low temperatures. Appl Therm Eng 128:1382–1390

98. Wang LW, Metcalf SJ, Critoph RE, Tamainot-Telto Z, Thorpe R (2013) Two types of natural graphite host matrix for composite activated carbon adsorbents. Appl Therm Eng 50:1652–1657

99. Wang R, Wang L, Wu J (2014) Sorption refrigeration technology: theory and application. Wiley

100. Kiplagat JK, Wang RZ, Oliveira RG, Li TX, Liang M (2013) Experimental study on the effects of the operation conditions on the performance of a chemisorption air conditioner powered by low grade heat. Appl Energy 103:571–580

101. Zhou ZS, Wang LW, Jiang L, Gao P, Wang RZ (2016) Non-equilibrium sorption performances for composite sorbents of chlorides—ammonia working pairs for refrigeration. Int J Refrig 65:60–68

102. Lee C (2016) Modeling urea-selective catalyst reduction with vanadium catalyst based on NH$_3$ temperature programming desorption experiment. Fuel 173:155–163

103. Meng B, Zhao Z, Chen Y, Wang X, Li Y, Qiu J (2014) Low-temperature synthesis of Mn-based mixed metal oxides with novel fluffy structures as efficient catalysts for selective reduction of nitrogen oxides by ammonia. Chem Commun 50:12396

104. Reay DA, Kew PA, McGlen RJ (2013) Heat pipes: theory, design and applications, 6th edn. Elsevier, Whitley Bay, United Kingdom, pp 1–251

105. Jafari D et al (2016) Two-phase closed thermosyphons: a review of studies and solar applications. Renew Sustain Energy Rev 53:575–593

106. Tang Y et al (2013) Effect of fabrication parameters on capillary performance of composite wicks for two-phase heat transfer devices. Energy Convers Manag 66:66–76

107. Jiang L et al (2014) Thermal performance of a novel porous crack composite wick heat pipe. Energy Convers Manag 81:10–18

108. Brahim T, Dhaou MH, Jemni A (2014) Theoretical and experimental investigation of plate screen mesh heat pipe solar collector. Energy Convers Manag 87:428–438

109. Hakeem MA, Kamil M, Arman I (2008) Prediction of temperature profiles using artificial neural networks in a vertical thermosiphon re-boiler. Appl Therm Eng 28:1572–1579

110. Abreu SL, Colle S (2004) An experimental study of two-phase closed thermosyphons for compact solar domestic hot-water systems. Sol Energy 76:141–145

111. Sundaram AS, Bhaskaran A (2011) Thermal modeling of thermosyphon integrated heat sink for CPU cooling. J Electron Cool Therm Control 1:15–21

112. El-Genk MS, Saber HH (1997) Flooding limit in closed, two-phase flow thermosyphons. Int J Heat Mass Transf 40(9):2147–2164

113. Nguyen-Chi H, Groll M (1981) Entrainment or flooding limit in a closed two-phase thermosyphon. J Heat Recover Syst 1(4):275–286

114. Shatto DP, Besly JA, Peterson GP (1997) Visualization study of flooding and entrainment in a closed two-phase thermosyphon. J Thermophys Heat Transf 11(4):579–581

115. Yu N, Wang RZ, Wang LW (2013) Sorption thermal storage for solar energy. Prog Energy Combust Sci 39:489–514

116. Flueckiger SM, Volle F, Garimella SV et al (2012) Thermodynamic and kinetic investigation of a chemical reaction-based miniature heat pump. Energy Convers Manag 64(64):222–231

117. Jiang L, Wang LW, Zhou ZS et al (2016) Investigation on non-equilibrium performance of composite adsorbent for resorption refrigeration. Energy Convers Manag 119:67–74

118. Vasiliev L, Vasiliev L (2005) Sorption heat pipe-a new thermal control device for space and ground application. Int J Heat Mass Transf 48(12):2464–2472

119. Vasiliev L, Vasiliev L (2004) The sorption heat pipe-a new device for thermal control and active cooling. Superlattices Microstruct 35(3–6):485–495
120. Yu Y, Wang LW, Jiang L, Gao P, Wang RZ (2017) The feasibility of solid sorption heat pipe for heat transfer. Energy Convers Manag 138:148–155
121. Yu Y, Wang LW, An GL (2018) Experimental study on sorption and heat transfer performance of NaBr-NH3 for solid sorption heat pipe. Int J Heat Mass Transf 117:125–131
122. Jasvanth VS et al (2017) Design and testing of an ammonia loop heat pipe. Appl Therm Eng 111:1655–1663
123. Song H et al (2016) Experimental study of an ammonia loop heat pipe with a flat plate evaporator. Int J Heat Mass Transf 102:1050–1055
124. Junior JB, Vlassov VV, Genaro G, Guedes UTV (2015) Dynamic test method to determine the capillary limit of axially grooved heat pipes. Exp Thermal Fluid Sci 60:290–298
125. Smitka M, Kolková Z, Nemec P, Malcho M (2014) Impact of the amount of working fluid in loop heat pipe to remove waste heat from electronic component. EPJ Web Conf 67:02109
126. Vantúch M, Malcho M (2014) Influence of structural design condensing part of NH3 heat pipe to heat transfer. EPJ Web Conf 67:02123
127. Xue ZH, Qu W (2014) Experimental study on effect of inclination angles to ammonia pulsating heat pipe. Chin J Aeronaut 27:1122–1127

Chapter 6
Solid Sorption Cycle for Energy Storage, Electricity Generation and Cogeneration

Abstract In this chapter, composite sorbents in heat and refrigeration cogeneration cycle, refrigeration and electricity cogeneration cycle based on sorption or resorption technique, resorption power generation cycle for energy storage, electricity generation and cogeneration are presented.

Keywords Sorption · Resorption · Heat · Refrigeration · Energy storage · Electricity · Cogeneration

6.1 Solid Sorption Cycle for Energy Storage

Thermal Energy Storage (TES) technology can eliminate the contradiction between energy supply and demand [1], and provides a promising method for the utilization and recovery of low-grade thermal energy such as geothermal resources, solar energy and industrial waste gas [2–5].

TES methods are generally divided into three categories: thermochemical, sensible and latent [6–8]. Sensible heat and latent heat storage are the most studied technologies in recent decades. For sensible heat storage, energy is stored by changing the temperature of materials such as rocks, water, aquifers, and bricks. The heat storage is directly proportional to the specific heat, mass and temperature difference of the heat storage medium [9].

Latent heat storage relies on the phase change process of paraffin, nitrate, fatty acid and other phase change materials (PCMs) at a constant temperature to charge and release thermal energy [10]. Compared with sensible heat storage, latent heat storage has higher energy storage density and smaller temperature difference [11]. However, both sensible heat storage and latent heat storage inevitably have the disadvantages of relatively low energy storage density, unacceptable energy loss, and not suitable for regional and seasonal energy storage [12].

Sensible heat storage has temperature fluctuations in the heat release process, which is difficult to meet the needs of end users. Phase change energy storage materials have the defects of sub-cooling and phase segregation during phase changing process. It is difficult for phase change materials to avoid the shortcomings of low

© Science Press 2021
L. Wang et al., *Property and Energy Conversion Technology of Solid Composite Sorbents*, Engineering Materials, https://doi.org/10.1007/978-981-33-6088-4_6

thermal conductivity [13] and severe performance deterioration during repeated heating and cooling cycles [14]. By contrast, thermochemical storage provides a feasible way to overcome the above shortcomings by weakening the mismatch between energy demand and supply as well as adjusting the space-time discrepancy [15, 16].

Sorption energy storage is a promising thermochemical storage method, because in desorption and sorption processes, the chemical potential between different solid-gas pairs is very large [7]. In the sorption process, energy is stored by breaking the chemical potential between the sorbent and sorbate. In addition, through the exergy and energy analysis of various studies, the sorption energy storage technology has been proven to be one of the most effective TES technologies, and it is also considered as a seasonal energy storage technology with negligible energy loss.

Sorption heat storage has attracted wide attention due to its high feasibility and energy density [17]. Korhammer et al. [18] proposed the design criteria and sorption characteristics of the composite sorbent based on $CaCl_2$ for thermochemical energy storage. Yu et al. [6] summarized different sorption heat storage technologies driven by solar energy, and compared the advantages and disadvantages of different solid-gas working pairs. Yan et al. [19] gave the selection criteria for chemical reactions in thermochemical storage and summarized some promising reversible reactions. Li et al. [20] analyzed sorption thermal battery for solar heating and cooling energy storage and heat transformer based on solid-gas reaction process. Results indicated that the highest heat and cold energy storage density can reach 1600 and 720 kJ/kg using working pair of $SrCl_2$-NH_3. Haije et al. [21] investigated the property of a $LiCl$-$MgCl_2$-NH_3 sorption heat storage and experimentally achieved an energy density of 222 W/kg and COP of 0.11, respectively. The researches on sorption heat storage mainly focus on new sorbents, advanced discharging and charging cycles and optimized configuration of heat-exchange reactor [22].

Compared with sorption [23], resorption refrigeration has a higher cooling capacity and COP, because the cooling capacity is generated by the reaction [24] instead of the latent heat of the refrigerant. At the same time, due to the low content of liquid ammonia in the system, it has the advantages of a safe structure. In this section, a novel resorption thermal energy storage (RTES) system using the working pair of $MnCl_2$-$CaCl_2$-NH_3 is established to study the performance of heat and refrigeration cogeneration [25], and performance analysis is introduced to fully understand the system.

6.1.1   Establishment of the Heat and Refrigeration Cogeneration Cycle

When the equilibrium pressure is 1.5 MPa, the sorbents driven by temperatures higher than 150 °C, between 90 and 150 °C, and temperatures lower than 90 °C are called high temperature sorbents (HTSs such as $NiCl_2$, $MnCl_2$, $FeCl_3$), medium temperature sorbents (MTS such as $SrCl_2$, $CaCl_2$) and low temperature sorbents (LTS such as NaBr, NH_4Cl, $BaCl_2$), respectively [26]. Considering the cyclic sorption capacity and matching the temperature of the heat source, $MnCl_2$ and $CaCl_2$ are selected as the high temperature sorbent and medium temperature sorbent of the RTES system according to the above criteria.

The operation steps of the resorption system for heat and refrigeration cogeneration are based on the energy conversion between the bond energy of the sorption potential and the thermal energy in the solid-gas sorption process. The Claperon diagram of the thermodynamic cycle is shown in Fig. 6.1a, which mainly involves energy storage and the cogeneration process of heat and cooling.

The thermodynamic cycle can be broadly divided into two processes: discharging (cold and heat cogeneration) and charging (energy storage) process (Fig. 6.1b and c). During charging (A–B) process which consists of desorption in HTS reactor and sorption in MTS reactor (Fig. 6.1b), the high temperature sorbent stores thermal energy in the form of bond energy, which is generated by a desorption process at high temperature. When the high temperature sorbent reactor is heated to a certain temperature (point A in Fig. 6.1a), the ammonia attached to the $MnCl_2$ sorbent begins to escape. Then, gaseous ammonia flows through the pipeline under the pressure difference and is sorbed by the MTS reactor at a lower temperature (point B in Fig. 6.1a). Simultaneously, the heat of sorption is released into the environment. It is worth emphasizing that some additional sensible heat is added to the system to heat the solid sorbent, metals and other components. After the energy storage phase, close the ammonia valve between the HTS and MTS reactors to avoid chemical reactions by separating sorbent and sorbate (Fig. 6.1b and c). Once heating or cooling is required, the discharge process can be realized by connecting MTS and HTS reactors (Fig. 6.1c). In this process (C–D), the system performs the sorption process in the HTS reactor (point D in Fig. 6.1a) and desorption process in the MTS reactor (point C in Fig. 6.1a) at the same time. The energy storage is released in the form of cold energy and heat energy. The MTS reactor provides a cooling effect at point C, and

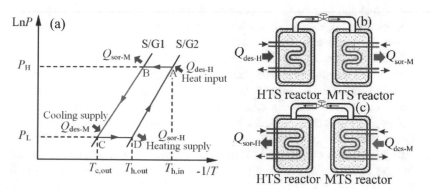

Fig. 6.1 Schematic diagram of resorption system. **a** Clapeyron diagram of the cycle; **b** Charging process; **c** Discharging process [25]

(a)   (b)

(1) PCM energy storage tanks; (2) HTS reactor; (3) heat exchanger between oil and water for

cooling the oil tank; (4) LTS reactor; (5) cooling tower; (6) safety valve; (7) pressure gauge; (8)

inlet ammonia valve for power generation; (9) main valve between HTS and LTS reactor; (10)

outlet ammonia valve for power generation

Fig. 6.2 Picture of cogeneration system. **a** Resorption refrigeration part; **b** Auxiliary equipment [25]

the HTS reactor provides a heating effect at point D according to the needs of the end user. Both sensible heat and sorption heat can be used to provide useful heat output.

Figure 6.2a shows the main part of the resorption system. Six shell-and-tube vessels form two sorption reactors, i.e. high temperature sorbent and medium temperature sorbent reactor. In the fin gaps of the fin tubes in the MTS and HTS reactors, 3.9 kg $CaCl_2$ and 4.8 kg $MnCl_2$ composited with expanded natural graphite treated with sulfuric acid (ENG-TSA) are compressed respectively. There are two PCM energy storage tanks in the resorption system, marked with "1" in Fig. 6.2a. The upper energy storage tank contains 40% $NaNO_2$, 53% KNO_3 and 7% $NaNO_3$ ternary mixed molten nitrate salt to storage the excess heat of heating oil during charging phase, another energy storage tank is filled with tetrabutyl ammonium bromide to store cold to achieve the purpose of continuous refrigeration. Figure 6.3 shows the structure of the HTS/MTS reactors in detail. The length, outer diameter and fin height of each fin tube are 1 m, 25 mm and 32 mm respectively. After the composite sorbent is squeezed into the fin gaps of the fin tubes to reach a certain density, a stainless steel mesh is coated on the surface of the fin tube to effectively prevent the agglomeration of the sorbent during the sorption process. Each vessel of the HTS/MTS reactors includes six finned tubes. For the development of the sorbent [27], the granular ENG-TSA is firstly dried in an oven at 150 °C for 2 h to eliminate the mass error caused by impure gas and moisture. Secondly, the alkali halide and water are mixed to form homogeneous solution. Then put the dry ENG-TSA into the solution and stir the mixture. After that, put the prepared mixture in an oven and place it at 120 °C for 12 h. Due to the presence of crystal water in some alkali metal halides, it should be

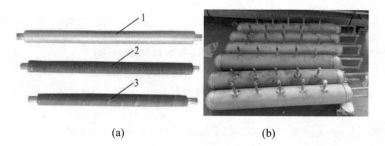

(a)   (b)

Fig. 6.3 Structure of MTS/HTS reactors. **a** Finned-tube: 1-bare finned tube; 2-finned tube with sorbent; 3-finned tube covered with coated screen; **b** Containers of MTS/HTS reactors [25]

placed in an oven for 12 h at a high temperature of 220 °C to remove the crystal water.

Due to its porous structure, ENG-TSA can improve the sorption and heat transfer performance. At the same time, it can also improve mass transfer by significantly preventing agglomeration [28]. The temperature of the high temperature sorbent reactor is controlled by a cooling tower and an oil tank, while the temperature of the medium sorbent reactor is regulated by a water thermostat. Figure 6.2b shows a picture of the RTES system and its auxiliary equipment, cooling tower, oil tank, thermostat and RTES components.

In order to fully understand the resorption system, different charging, cooling and discharging temperatures are studied. The cooling temperature is controlled at 5, 10, and 15 °C to simulate air-conditioning conditions. The discharging temperature ranges from 30 to 50 °C with an increment of 5 °C and the charging temperature of oil ranges from 130 to 160 °C.

6.1.2 Performance Analysis

Some parameters for evaluating the performance of the resorption system are given in the article. The formula for calculating the heat input during the charging phase is:

$$Q_{h,in} = \int_0^{t_{h,in}} c_o \cdot m_o \cdot (T_{HTS,i} - T_{HTS,o}) dt \qquad (6.1)$$

where $t_{h,\,in}$ is the charging time, $T_{HTS,\,i}$ and $T_{HTS,\,o}$ are inlet and outlet temperature of HTS reactor, $Q_{h,\,in}$ is the charging heat provided by heating oil, c_o and m_o are specific heat and mass flow rate of heating oil.

$$Q_{h,sen} = \int_0^{t_{coo}} c_o \cdot m_o \cdot (T_{HTS,o} - T_{HTS,i}) dt \qquad (6.2)$$

where $Q_{h, sen}$ is the sensible heat output of HTS reactor during cooling phase and t_{coo} is the cooling time.

The heat output during discharging phase is calculated by:

$$Q_{h,rel} = \int_0^{t_{h,rel}} c_o \cdot m_o \cdot (T_{HTS,o} - T_{HTS,i}) dt \tag{6.3}$$

where $Q_{h, rel}$, $t_{h, rel}$ are the discharging heat and time respectively.

$$Q_{h,out} = Q_{h,sen} + Q_{h,rel} \tag{6.4}$$

Heat storage density is defined as:

$$HSD = \frac{Q_{h,out}}{m_{so}} \tag{6.5}$$

where m_{so} is the total mass of sorbents i.e. $CaCl_2$ and $MnCl_2$.

Heat storage efficiency and the ratio of latent heat to sensible heat are defined as:

$$COP_h = \frac{Q_{h,out}}{Q_{h,in}} \quad \varepsilon = \frac{Q_{h,sen}}{Q_{h,rel}} = \frac{Q_{h,sen}}{Q_{h,lat}} \tag{6.6}$$

where $Q_{h, lat}$ and $Q_{h, sen}$ are the latent and sensible heat output respectively.

The refrigerating capacity is calculated by:

$$Q_{ref} = \int_0^{t_{ref}} c_W \cdot m_W \cdot (T_{MTS,i} - T_{MTS,o}) dt \tag{6.7}$$

where t_{ref} is refrigerating time, $T_{MTS, i}$, $T_{MTS, o}$ are inlet and outlet temperature of MTS reactor, c_w, m_w are specific heat and mass flow rate of cooling water.

Heat and refrigeration exergy are calculated by:

$$E_{h,in} = Q_{h,in} \cdot (1 - \frac{T_0}{T_{chr}}) \tag{6.8}$$

$$E_{heat} = Q_{h,out} \cdot (1 - \frac{T_0}{T_{h,ave}}) \tag{6.9}$$

$$E_{cold} = Q_{ref} \cdot (\frac{T_0}{T_{c,ave}} - 1) \tag{6.10}$$

where T_0 is the ambient temperature, $T_{c, ave}$ is the average temperature of chilled water, T_{chr} and $T_{h, ave}$ are the average charging and discharging temperature.

The coefficient of performance and specific cooling power per kilogram sorbent are expressed as:

$$\mathrm{COP_c} = \frac{Q_{\mathrm{ref}}}{Q_{\mathrm{h,in}}}; \quad \mathrm{SCP} = \frac{Q_{\mathrm{ref}}}{M_{\mathrm{ad}} \cdot t_{\mathrm{cyc}}} \tag{6.11}$$

where t_{cyc} is cycle time.

Total energy efficiency and exergy efficiency:

$$\eta_{\mathrm{ene}} = \frac{Q_{\mathrm{h,out}} + Q_{\mathrm{ref}}}{Q_{\mathrm{h,in}}}; \quad \eta_{\mathrm{exe}} = \frac{E_{\mathrm{heat}} + E_{\mathrm{cold}}}{E_{\mathrm{h,in}}} \tag{6.12}$$

6.1.3 Results and Discussions

6.1.3.1 Heat Storage Performance

Under different charging temperatures, the pressure and temperature change trends of the high temperature sorbent reactor during the charging stage are similar. Figure 6.4a shows the temperature changes of different parts of the HTS reactor during charging phase when the heat source temperature is 130 °C. Figure 6.4b shows the pressure change of the HTS reactor, which represents the pressure of the MTS reactor. Since the valve between the MTS and HTS reactors remains open during the energy storage phase, it further characterizes the pressure of the system.

As seen in Fig. 6.4a, the charging phase of about 45 min consists of several phases which fail to emerge in the picture. In the first stage of about 15 min, the temperature of the HTS reactor increases gradually because the heat consumed is the energy to heat the reactor to a certain desorption temperature, and the desorption temperature is related to the mass ratio of the sorbent and auxiliary metal components, temperature

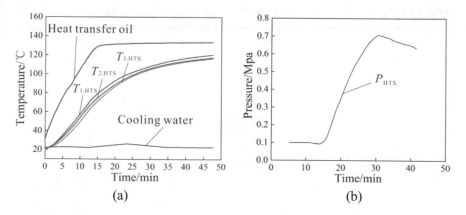

Fig. 6.4 Temperature and pressure variations of HTS reactor during charging phase under the temperature 130 °C. **a** Temperature; **b** Pressure [25]

difference and heat loss. Within 15 to 35 min, the heat sorbed in this stage is used to provide desorption heat of the composite sorbents in the HTS reactor, so that the temperature of the reactor rises steadily. With the effect of thermal resistance along the route of heating oil, $T_{3,\ HTS}$ is lower than $T_{2,\ HTS}$ and $T_{2,\ HTS}$ is lower than $T_{1,\ HTS}$. Meanwhile, the MTS reactor has undergone a sorption process, and the heat released during this process explains the change in the temperature of the cooling water at the outlet of the MTS reactor. In the last 10 min, due to the influence of the heat transfer temperature difference, the temperature of the HTS reactor shows a stable trend, which also affects the termination of the desorption process of the HTS reactor. The pressure remains basically unchanged at 0.1 MPa in the first 15 min, then rises sharply from 0.1 to 0.7 MPa in the next 15 min, and gradually decreases in the last 15 min, which indirectly prove the above stages. It is worth emphasized that when the system reaches a peak pressure of 0.7 mPa at about 30 min, the sorption process in the MTS reactor and the desorption process in the HTS reactor have the same reaction rate. The main ammonia valve is closed in order to avoid the chemical reaction to achieve energy storage at the end of charging phase. Subsequently, the HTS reactor is cooled to a discharge temperature of 30–50 °C, while the MTS reactor is controlled to the design temperature to simulate air-conditioning conditions. When heating or cooling is required, the HTS reactor is connected to the MTS reactor to meet the demand by performing the aforementioned sorption and desorption process.

A complete cycle consisting of three phases: discharging phase, charging phase, and cooling phase, takes about 90 min, and each phase accounts for 30 min, 45 min, and 15 min, respectively. Due to the use of high-power cooling tower and an oil-water heat exchanger to cool the HTS reactor, the cooling phase occupies a small part of the entire cycle time. After the cooling phase is over, the ammonia valve between the MTS and HTS reactor is opened to release the stored energy, and the HTS reactor is adjusted to 30–50 °C. When the MTS reactor is 15 °C, the theoretical equilibrium temperature of the HTS reactor is 64 °C, that is, the lower the discharge temperature is, the more obvious the effect is. In order to give the impression of resorption system during the discharging phase, the maximum outlet temperature increment of the HTS reactor is introduced to describe the heat storage effect, which represents the maximum temperature difference between the discharge temperature and the circulating oil outlet temperature. Table 6.1 shows the largest outlet temperature

Table 6.1 The largest outlet temperature increment of HTS reactor during discharging phase [25]

The largest outlet temperature increment/°C		Charging temperature/°C			
		130	140	150	160
Discharging temperature/°C	50	4	6	9	10
	45	8	10	11	12
	40	13	14	15	16
	35	14	15	16	17
	30	17	18	18	20

increment under different discharging and charging temperature with cooling temperature of 15 °C. Under various working conditions, the outlet temperature increment is between 4 and 20 °C. The results show that as the discharge temperature decreases and the charging temperature increases, the maximum temperature rise increases. The temperature change during the discharge stage also shows a similar trend, that is, it rises sharply at the beginning and then gradually decreases. This is mainly because the desorption process in the MTS reactor has a faster reaction rate than the sorption process in the HTS reactor. The charging temperature has an important effect on the heat releasing. The higher the charging temperature, the faster the reaction rate. Similarly, a lower discharge temperature means a larger temperature difference, so a higher outlet temperature increase can be expected.

Energy storage density is usually used to evaluate the performance of TES and can introduce it into the sorption system according to the amount of sorbent filled, as shown in Fig. 6.5. Figure 6.5a shows the energy density at different charging and discharging temperature with a cooling temperature of 15 °C. The energy density range is 912–1706 kJ/kg. When the discharging temperature is 30 °C and the charging temperature is 160 °C, the energy density is the highest. It can be seen from the figure that the energy density increases with the decrease of the discharging temperature and the increase of the charging temperature, which is consistent with the rising trend of the outlet temperature shown in Table 6.1. Figure 6.5b shows the energy density at different charging and cooling temperature when the discharging temperature is set to 30 °C. The energy density is between 1159 and 1629 kJ/kg. When the charging and cooling temperatures are 155 °C and 15 °C respectively, the energy density is the highest. The energy density increases with the cooling and charging temperature, as shown in Fig. 6.5b. The corresponding theoretical discharging temperature increases with the increase of cooling temperature, which provides a higher temperature driving force for heat dissipation.

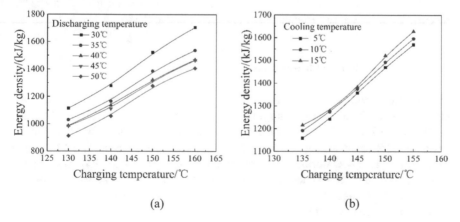

(a)   (b)

Fig. 6.5 Heat storage density for resorption system. **a** Density with discharging temperature; **b** Density with cooling temperature [25]

Fig. 6.6 Heat storage efficiency and the ratio of sensible heat to latent heat under different charging temperature [25]

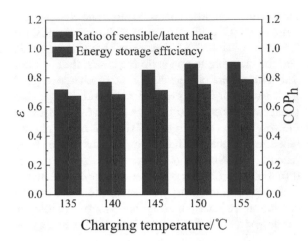

Figure 6.6 describes the energy storage efficiency and the ratio of sensible heat to latent heat at different charging temperatures of 135–155 °C when the cooling and discharging are the same temperature of 15 °C and 30 °C, respectively. The results show that as the charge temperature increases, the ratio of sensible heat to latent heat and the energy storage efficiency increase. The ratio of sensible heat to latent heat is 0.72–0.90, and the energy storage efficiency is 0.67 to 0.79. When the charging temperature is lower than 150 °C, the ratio of sensible heat to latent heat increases sharply. This is mainly because the HTS reactor only partially desorbs during the charging phase, resulting in little ammonia gas sorbed by the HTS during the discharge phase. When the charging temperature is higher than 150 °C, the cyclic sorption capacity determined by the charging temperature is almost unchanged, and the energy storage efficiency changes little. This phenomenon means that HTS can completely desorb during the charging stage, and the heat input is mainly used to provide sensible heat. When the charging temperature increases from 135 to 155 °C, the corresponding exergy efficiency is between 0.28 and 0.34.

6.1.3.2 Performance of Resorption Refrigeration

The changing trend of different refrigeration temperature with the charging temperature is similar. When the discharging temperature is 30 °C, the air-conditioning condition of 15 °C is chosen as the example. Figure 6.7 shows the average cooling power at different charging temperatures when the cooling time is 20 min. The average cooling power during the discharging phase is 0.59–1.07 kW, and reaches the maximum value at the charging temperature of 160 °C. If the average cycle time is 90 min, the average cooling power will drop to 238 W. The SCP and COP of the resorption system at different charging temperature are shown in Fig. 6.8, and their ranges are 15–27.33 W/kg and 0.05–0.07, respectively. As shown in the figure, both COP and SCP increase with the increase of charging temperature, and the maximum

Fig. 6.7 Average cooling power under different charging temperature during discharging phase [25]

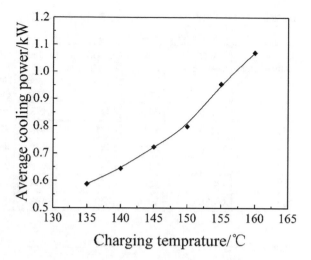

Fig. 6.8 Refrigeration performance of resorption system under different charging temperature [25]

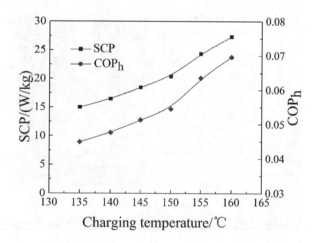

appears when the charging temperature is 160 °C. Since the heat input is quite large and most of the cooling power is consumed by sensible heat, the COP and SCP values are small.

6.1.3.3 Heat and Refrigeration Cogeneration

Table 6.2 shows the energy balance of the RTES system when the discharge temperature is 35 °C and the cooling temperature is 15 °C. Energy and exergy efficiency derived from Eq. 6.12 are calculated to evaluate the resorption system for refrigeration and heat cogeneration. Figure 6.9 describes the energy and exergy efficiency of the resorption system at different charging temperatures. The energy efficiency

Table 6.2 Energy and exergy balances for RTES system [25]

Charging temperature (°C)	Cooling temperature = 15 °C; Discharging temperature = 35 °C						
	Total heat input (kJ)	Total exergy input (KJ)	Sensible heat output (KJ)	Latent heat output (KJ)	Refrig capacity output (KJ)	Total energy loss (KJ)	Total exergy loss (KJ)
135	15710	4426	4419	6158	705	4427	3152
140	16243	4717	4841	6288	773	4341	3333
145	16908	5054	5544	6494	868	4001	3498
150	17574	5399	6248	6998	957	3370	3638
155	18016	5680	6730	7438	1144	2704	3752
160	18458	5965	7213	7628	1284	2332	3888

Fig. 6.9 Energy and exergy efficiency for the resorption system under different charging temperature [25]

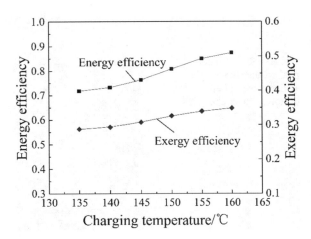

is between 0.72 and 0.87, and it increases as the charging temperature increases, as shown in the Fig. 6.9, When the charge temperature increases from 135 to 160 °C, the exergy efficiency rises steadily from 0.29 to 0.35.

Heat storage density and the total energy efficiency are two main parameters involved in the RTES system performance evaluation. The relative tolerances are calculated as follows:

$$\left|\frac{dHSD}{HSD}\right| \leq 2\left|\frac{dm_{so}}{m_{so}}\right| + \left|\frac{dQ_{h,sen}}{Q_{h,sen}}\right| + \left|\frac{dQ_{h,rel}}{Q_{h,rel}}\right| = 2\left|\frac{dm_{so}}{m_{so}}\right| + 2\left|\frac{dm_o}{m_o}\right| + 2\left|\frac{d\Delta T}{\Delta T}\right| \tag{6.13}$$

$$\left|\frac{d\eta_{ene}}{\eta_{ene}}\right| \leq 3\left|\frac{dQ_{h,in}}{Q_{h,in}}\right| + \left|\frac{dQ_{h,sen}}{Q_{h,sen}}\right| + \left|\frac{dQ_{h,rel}}{Q_{h,rel}}\right| + \left|\frac{dQ_{ref}}{Q_{ref}}\right| = 5\left|\frac{dm_o}{m_o}\right| + \left|\frac{dm_w}{m_w}\right| + 6\left|\frac{d\Delta T}{\Delta T}\right| \tag{6.14}$$

The maximum relative tolerances of heat storage density and energy efficiency are 4% and 9%, respectively.

6.2 Solid Sorption Cycle for Electricity and Refrigeration Cogeneration

Sorption refrigeration technology is an alternative refrigeration technology that recovers low-grade heat energy. In the desorption process of the sorption cycle, the working fluid is released under high pressure and high temperature. If a generator is added, the turbine/expander can be driven to generate electrical or mechanical energy. In the sorption process, the cooling effect is produced by the evaporation of the working fluid. Since cooling power generation and electricity generation are relatively separated in different half cycles, there is a potential for high cooling efficiency in cogeneration [29]. Although the sorption method has many advantages, the sorption is unstable and discontinuous output. If combined with the expansion process, the evolution of chemisorption will affect the stability of power generation; Vice versa, the expansion process will weaken the sorption cycle performance.

In order to find a solution, a sorption cogeneration prototype system is designed [30]. The system consists of two sorption units to overcome the intermittent phenomenon, where a mixture of activated carbon and $CaCl_2$ is used as the sorbent, ammonia is used as the refrigerant, and an oil-free scroll expander is chosen for converting low-grade heat into electricity [31–34].

6.2.1 Cogeneration Principle and Theoretical Analysis

The two sorption devices work together to achieve continuous cogeneration. Each sorption unit consists of a sorption bed, a condenser and an evaporator, as shown in Fig. 6.10. The water container and boiler are used to store heat-sink water and generate heat source steam respectively. The desorbed refrigerant enters the expander to convert heat energy into mechanical energy, while the sorbent draws the evaporated refrigerant out of the evaporator to produce a cooling effect. The principle of cogeneration is as follows:

(1) Electric power generation. As shown in Fig. 6.10a, Ads 1 (sorbent bed 1) is heated by the hot steam from the boiler through V1, and then returns to the boiler through V3. Because the thermodynamic state of the sorbent in Ads 1 deviates from equilibrium, the high-pressure and high-temperature refrigerant is released to the expander through the one-way valve and V5. If a generator is installed, it will generate power output. The exhausted refrigerant flows through V6 and the Con 1 (condenser 1), and ends up in the Evp 1 (evaporator 1).

(2) Cooling power generation. The internal circulation heat sink maintains a certain temperature and dissipates heat through the heat exchanger connected to the external cooling tower. In order to make the whole system compact and simplify the water pipeline, the internal heat sink maintains a constant circulation route, flowing through Con 2, Con 1 and Ads 2 (or Ads 1, depending on the cycles), and finally return to the water container. Once Ads 2 is cooled, it begins to sorb

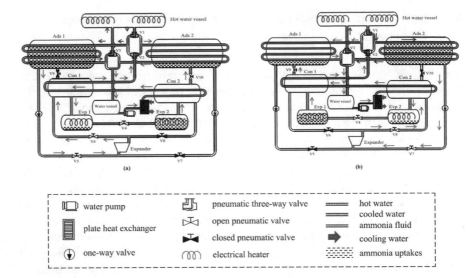

Fig. 6.10 The schematic of Ad-Cogen system. **a** Sorbent bed 1 side electricity generation and the sorbent bed 2 side refrigeration; **b** Sorbent bed 2 side electricity generation and the sorbent bed 1 side refrigeration [30]

the refrigerant from Evp 2, where it captures the heat of vaporization from the surrounding environment, thereby producing a cooling effect.

(3) Alternate operation. As shown in (1) and (2) above, Ads 1 performs desorption, while Ads2 simultaneously performs sorption. Then, the two sorbent beds are exchanged, that is, three-way valves V1, V2, and V3 are switched, V5, V6, and V10 are closed, and V7, V8, and V9 valves are opened, as shown in Fig. 6.10b. Ads 1 will be disconnected from expander and sorb refrigerant from Evp 1 to generate cooling, while Ads 2 will be integrated with expander and change into desorption to supply electric power generation.

(4) Repeat the above three steps to achieve continuous cogeneration. A mass recovery process is introduced in the switching operation to improve the capacity and performance of the circulating refrigerant. In other words, when Ads 1 desorbs refrigerant vapor and becomes unsaturated at a higher pressure, and Ads 2 sorbs vapor and approaches saturation at a lower pressure, the mass recovery could be carried out by opening the valve between Ads 1 and Ads 2 for a short time period. Therefore, the pressure in these two beds will become the same, and the sudden pressure swing causes: Ads 1 undergoes a further desorption and Ads 2 experiences an extended sorption, leading to the increment in the cyclic amount of refrigerant. This approach has been proven effective in many researches [35–37].

Figure 6.11 shows the ideal working principle of the Claperon diagram with $CaCl_2$ as the reactive sorbent. A to B in Fig. 6.11 represent the heating process of the sorbent. The pressure and temperature of the sorbent gradually increase along the equilibrium

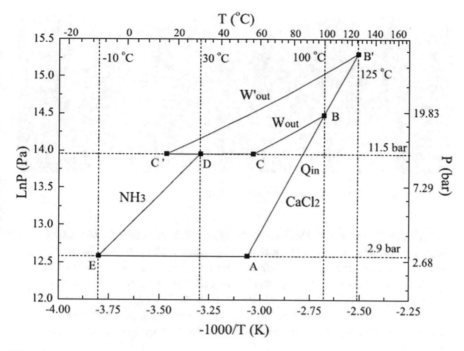

Fig. 6.11 Claperon diagram of CaCl$_2$-NH$_3$ sorption cogeneration process [30]

line until the pressure (point B or B') is higher than the saturation pressure of ammonia corresponding to the heat sink temperature (point D). The process B to C represents the isentropic expansion process. In order to increase the output power, the heating temperature can be adjusted higher to achieve an increase (compare B'–C' and B–C). The process from B to C is followed by refrigerant condensation, from C to D and then to E. Then, when the sorbent is sufficiently cooled and the pressure drops below the pressure of the saturated refrigerant at low temperature, the E to A process occurs.

6.2.2 Experiment Set-Up

Figure 6.12 are the pictures of the cogeneration prototype system. Four shell-and-tube vessels are used as two sorbent beds. Each vessel contains two connected vessels, and each vessel is wrapped with 6 finned tubes. There are 17.5 kg CaCl$_2$ impregnated in activated carbon with the mass ratio of 4:1 to form the composite sorbent which is squeezed into the fin gaps of the finned pipes, while the heat exchange fluid flowed through the finned tubes and the refrigerant occupied the rest space inside the shell. Due to its additional physical adsorption capacity and porous structure, activated carbon can effectively improve the sorption efficiency and mass transfer [28, 38,

Fig. 6.12 Photos of the sorption cogeneration prototype [30]

39]. The condenser is also shell-and-tube type, the heat-sink is in the tube and the refrigerant is in the shell. The evaporator is equipped with an electric heater to record the electric energy consumed to heat the liquid refrigerant. According to the heat balance theory, this electric energy can be regarded as an effective cooling capacity. This design eliminates some possible interference and enhances the compactness of the system. The system is charged with 28 kg of ammonia, which is determined by the amount of $CaCl_2$ plus the amount of the half-full evaporator to avoid the heater from dry heat.

It adopts an oil-free scroll expander with a rated output power of 1 kW and is connected to an AC generator through magnetic coupling. The expansion ratio is 3.5:1. The magnetic coupling eliminates refrigerant leakage. Pressure sensors are installed in the inlet and outlet of the expander, sorbent bed and evaporator. A vortex flowmeter is installed in the straight pipe section at the front of the expander. The power meter records the real-time output of current, voltage and power. A pocket tachometer is used to measure the rotation speed of the expander.

6.2.3   Results and Discussions

6.2.3.1   Sorption Refrigeration

Figure 6.13a shows the temperature in the condenser/evaporator during the desorption and sorption of the two sets of sorbent beds during the refrigeration test. Figure 6.13b shows the corresponding pressure. The comparative analysis between tests with mass recovery and without mass recovery is based on two semi-stable cycles. The sorption capacity is strengthened and more sufficient desorption benefits from mass recovery. As a result, more cooling power and higher COP are obtained, which are increased by approximately 53% and 43%, respectively, compared with non-mass recovery cycles, as shown in Table 6.3. The advantage of a more complete desorption and sorption process is a wider range of pressure and temperature changes.

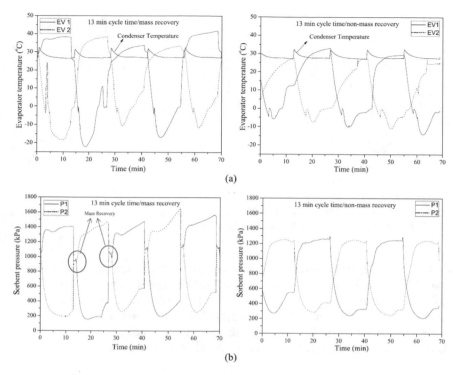

Fig. 6.13 Comparison of the sorption refrigeration cycles with and without mass recovery with the heat source at 125 °C, heat sink at 27 °C and cycle time of 13 min. **a** evaporation temperature; **b** system pressure [30]

Table 6.3 Refrigeration performance of sorption unit with and without mass recovery, at 13 or 26 min cycles [30]

	Cycle time (min)	Average heating temperature (°C)	Average cooling temperature (°C)	Average cooling power (W)	Cooling capacity (kJ/kg)	COP
MR	13	125	−2.50	3190	284	0.20
Non-MR	13	126	−0.02	2090	186	0.14
MR	26	124	−0.80	2410	430	0.14

The maximum pressure in the mass recovery cycle desorption process is 2 to 4 bar higher than that without mass recovery.

Figures 6.13 and 6.14 compare the performance under different cycle time, distinguishing the fast reaction period from the inefficient part in the whole long process (26 min). The effective cooling process usually occurs in the first 12–15 min. Therefore, a cycle time of 13 min will be a suitable choice for the current prototype, as the data in Table 6.3 also proves this point. Although the cooling capacity of the 26 min

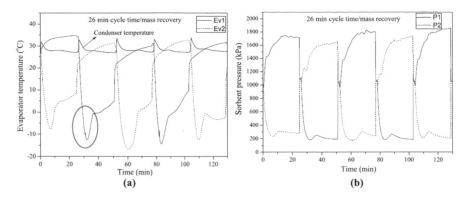

Fig. 6.14 Thermal variables of sorption refrigeration with mass recovery when the heat source at 125 °C, heat sink at 27 °C and cycle time of 26 min. [30]

cycle is 53% higher than that of the 13 min cycle (284 kJ/kg), the cooling power and COP of the 13 min cycle are 32% and 43% higher than those of the 26 min cycle, respectively. The low COP of the latter can be explained by the useless heat input during the additional 13 min of desorption. The pressure plateau at the end of each cycle in Fig. 6.14b also shows that during this period, most of the energy consumption is mainly wasted on sensible heat instead of reaction enthalpy.

6.2.3.2   Power Generation

The nitrogen is used for testing the performanc of scroll expander. The temperature is maintained at about 28–30 °C. The convective heat transfer on the outer surface of the expander is negligible and can be regarded as isentropic expansion. Figure 6.15 shows the rotation speed under different inlet pressures and steady-state performance of the scroll expander. The electricity power generation increases linearly from 21 to 530 W, and the speed increases from 1650 to 3300 rpm. At the same time, when the intake pressure increases from 4 to 11 bar, the exhaust temperature rises from 10 to 17 °C.

It can be seen from Fig. 6.16 that when the inlet pressure of the expander is low (about 4 bar), the gas can provide a stable supply. However, the air supply pressure vibrates as the required inlet pressure increases, resulting in unstable power output, as shown in Fig. 6.16b. Generally speaking, a higher inlet pressure can increase the rotating speed, so the expander tends to pump more working gas into the gas chamber, and power generation requires a larger flow rate. As shown in Fig. 6.16, the limitation of the gas supply of the nitrogen cylinder and the increase in the speed of the expander are the main reasons for the fluctuation of the pipe wall under high pressure. This also means that the proper flow rate has much impact on obtaining stable power output.

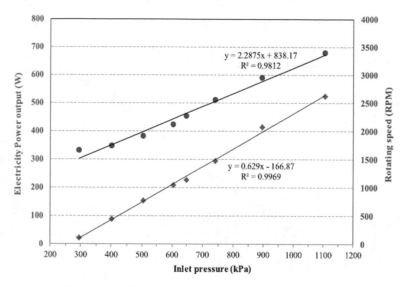

Fig. 6.15 Performance of the scroll expander with nitrogen as working gas at 28–30 under different inlet pressures [30] °C

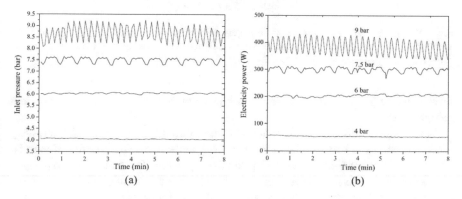

Fig. 6.16 Variations of inlet pressure and power generation with the nitrogen supplied from cylinder as working gas [30]

6.2.3.3   Cogeneration

As shown above, the heat source of cogeneration is 120–130 °C, but due to the heat transfer temperature difference, the temperature of ammonia released from the desorption bed is 85–95 °C and the pressure is 12–14 bar. The results show that in the cogeneration mode, the output electric power as a function of the pressure difference (Δp_{ex}) is almost the same as the test using nitrogen (4–11 bar, 28–30 °C), as shown in Fig. 6.17. This means that the cogeneration mode is comparable to the conventional simple power generation mode. In addition, the system can also be

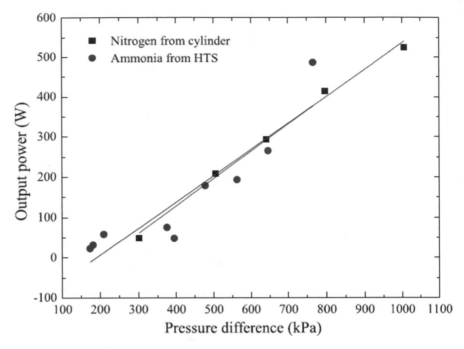

Fig. 6.17 Output electricity power both in cogeneration test with ammonia (12–14 bar and 85–95 °C at the inlet of expander) and in the expander-only test with nitrogen (4–11 bar and 28–30 °C at the inlet of expander) as working gas [30]

driven by low-grade thermal energy to achieve simultaneous power generation and refrigeration.

In the cogeneration test, it is found that the existence of the expander seriously hinders desorption, because the expander partly consumes Δp_{ad-con} that should be completely concentrated on the desorption driving force. The smaller the Δp_{ad-con}, the slower the desorption rate, which indicates that the smaller flow rate of the expander will cause the unstable rotation of the expander. In order to get as much power output as possible in the current situation, the method tried in this experiment is to keep V5 (or V7, depending on which one undergoes the desorption process) closed for a period of time and then release it. As shown in Fig. 6.18, this operation makes the flow rate appear as pulses. This allows the expander to generate electricity in several sections, because each time lasts a short time. The overall average value is 120 W, and the maximum output power is about 490 W. Due to the failure of continuous power generation, the total power generation is very small. The lowest temperature of the evaporator drops to 5.4 °C, and the cooling time is 3 min below 15 °C. Although operating the valve is not an ideal test method, it is worth exploring the potential of cogeneration. A key factor in establishing the prototype is to choose a small expander, which should match the size of the sorption chiller for stable operation. However, most off-the-shelf expander products are neither oil-free to avoid

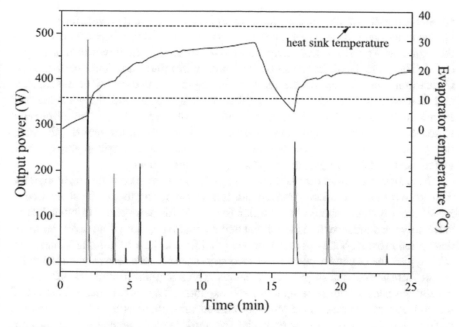

Fig. 6.18 Cogeneration performance of Ad-Cogen [30]

ammonia contamination by the refrigerant, nor are they small enough to couple with the sorption device at a given flow rate. On the other hand, the size of the sorption unit is limited by possible complexity and laboratory scale. Therefore, one system flawed during the beginning condition is the capability mismatch between the sorption unit and the scroll expander, which is not customized for this test specification. In addition, the maximum inlet pressure of the expander is limited to 13.8 bar, while the normal desorption pressure is at least 20 bar. Therefore, in order to prevent the desorption pressure from exceeding the limit of the expander, the heat source of this sorption test is controlled in a lower range, which causes the collateral losses of power generation and decreases the desorption effectiveness.

A contradictory fact in the cogeneration system is the contradiction between the stable input requirement for stable power generation and the peak curve of the desorption rate. As shown in Fig. 6.16, when the flow is not large enough, the pressure in the central chamber of the expander becomes wavy, especially in the high pressure area. In addition, by reducing the flow rate, this characteristic will gradually be exaggerated. Since the refrigerant is provided by chemisorption, there will be a period of time waiting for the desorbed refrigerant to accumulate to a sufficient amount, and then the next wave-like power generation can be expected. Moreover, according to the previous simulation study [29], the mutual restriction between downstream expansion and upstream desorption also plays an important role in unstable power generation. Since the thermodynamic requirements between expansion and desorption are similar, the capacity mismatch between the two may even worsen the situation. If they are connected in series, they will contradict each other. If the expander cannot let the desorbed gas pass in time, more refrigerant remaining in the sorbent bed will increase the constraint desorption pressure and prevent further work. On the other hand, the smaller the desorption rate, the slower the rotation speed of the expander. Therefore, less refrigerant passes through, which will form a vicious circle. The mutual constraints in cogeneration will affect individual performance. Furthermore, insufficient desorption will reduce the cycle capacity of the sorption unit, further deteriorate the overall performance, aggravate the vicious cycle, and even lead to non-repeatability and discontinuity of output. Due to anthropogenic factors and changes in the test environment, in the long-term test process, the sorption performance deteriorates, and the expander also has a certain degree of wear, which has a negative impact on the power generation capacity.

These problems make the mass recovery process an indispensable link to increase the capacity of the circulating refrigerant. It is worth noting that in the short time of mass recovery, there should be a barrier-free route for desorption. In other words, since only the condenser and the sorbent bed are involved, there is no expander in the desorption process, but a full development of additional desorption. The easiest way is to bypass the expander when mass recovery occurs. Mass recovery will equalize the pressure between all sorption components, which requires disconnection between the sorbent bed, condenser/evaporator and expander. The sorbent bed is heated and the condenser/evaporator is completely cooled until there is a sufficient pressure difference before and after the expander for proper power generation. Under non-ideal conditions, the refrigerant vapor received by the condenser/evaporator cannot be cooled in time to avoid excessive pressure at the outlet of the expander, which will affect the power generation process. In this case, mass recovery valve could be considered opening for a period of time, however at the same time the cooling effect will be significantly impaired. Therefore, at this point, the overall performance should be balanced by consciously operating the mass recovery on the premise of maximizing efficiency. For example, a short time offset operation time between the two sets of sorption cycles may be one of the potential solutions.

6.2.3.4 Potential Improved Performance

In order to solve the problems of the poor performance with the contradictory natures between the components and the existing design, the following attempts are needed:

(1) An expander that can withstand a higher inlet pressure (which should be equivalent to the desorption pressure) and upgrade the heat source to 125 °C. In this case, the exergy and (exergy) efficiencies of the 125 °C heat source are theoretically 0.25 and 0.46, respectively, which are 90% and 4.5% higher than the 100 °C heat source.

(2) Low-temperature heat sink to ensure that the saturation pressure of ammonia in the condenser is low, so that there is an enough Δp_{ex}. For example, at a given desorption pressure of approximately 13.8 bar, 350 W power output ideally requires approximately 7.5 bar Δp_{ex}, and the condensation temperature is 10 °C.

(3) Mass recovery should be carried out not only during the switching period, but also when the back pressure of the expander is not low enough. The mass

recovery offset operation for multiple sets of sorption units can be carefully arranged to obtain improvements.

(4) Energy storage devices such as super capacitors or batteries can be used to store unstable cogeneration output and regenerate electrical energy for stable use in the future.

The new design shown in Fig. 6.19 is a potential solution that can alleviate the mutual constraint between expander expansion and desorption in the reaction bed, and enable the operation of components fairly stable. A secondary condenser is added to the system, which can be installed between the expander and the sorbent bed. As shown in Fig. 6.19, the heat transfer fluid flowing through the secondary condenser can be the downstream heat-sink from the primary one. The condensation temperature and pressure on the secondary side should be higher than the condensation temperature and pressure on the primary side to provide a pressure difference to drive the expander. At the same time, the pressure in the secondary condenser should not be too high to avoid obstacles to desorption, otherwise there is a risk of no pressure difference between the secondary condenser and the sorbent bed. The key is to keep the temperature of the secondary condenser in a normal state. Therefore, the presence of this condenser divides the pressure difference between the primary condenser and the sorbent bed into two parts: one is the stable Δp_1 for desorption, and the other is the stable Δp_2 for expansion. By installing the condenser, the direct connection between the sorbent bed and the expander is avoided. At the same time, in order to alleviate the mismatch between the inlet flow required by the expander and the desorption rate, a buffer zone is designed. Therefore, as long as the pressure in the secondary condenser is maintained at an ideal value, desorption and expansion can be regarded as two independent processes. Although the maximum power output will be relatively lower than the original design, the operation of the entire system is

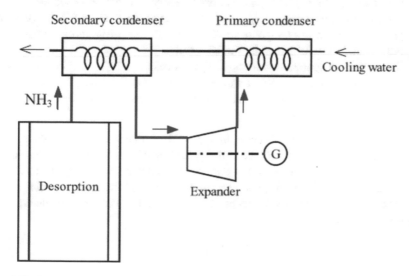

Fig. 6.19 Improved design to solve the mutual constraint between desorption and expansion connected in series [30]

expected to be continuous and stable. In theory, such a design can make the combined heat and power generation simply equivalent to the combined effect of desorption and expansion, making the performance of the system better than the initial design.

6.3   Resorption Cycle for Electricity and Refrigeration Cogeneration

In comparison with sorption technique, resorption technique employs the different halides that can match different reaction equilibrium temperatures to achieve the cascade utilization of energy. On basis of resorption cycle, two types of combined refrigeration and power generation cycle [40, 41] are proposed in this section.

6.3.1   Design and Performance Analysis of a Resorption Cogeneration System

6.3.1.1   Design of the Cycle

The combined resorption cycle for refrigeration and power generation is shown in Fig. 6.20. It mainly includes a low temperature sorbent (LTS) bed, a high temperature sorbent (HTS) bed, a pre-cooler, one turbine, a heating oil tank and two heat exchangers. Its working processes are presented in Fig. 6.20a and b:

(1) Electrical generation and sorption refrigeration process. In this process, the HTS bed is heated by the low-grade thermal energy, and the desorbed ammonia is expanded in the turbine. Expanded ammonia can provide refrigeration in the pre-cooler when the temperature is lower than the refrigeration temperature. After that, it will be sorbed by the LTS bed.

(2) Resorption refrigeration process. In this process, the HTS bed is cooled by the environmental medium. When the pressure is lower than the LTS bed, the HTS bed will sorb the refrigerant. The desorption process of LTS bed will produce refrigeration.

For the power generation process in Fig. 6.21, the heat source 2 heats the high temperature sorption bed firstly. Part of the heat provides the sensible heat for the sorbent and ammonia, and the other part provides the desorption heat of the high temperature sorption bed. After that, the ammonia steam is heated by the heat source 1, and then enters the turbine to generate electricity. The thermodynamic equation of power generation is:

$$W = \dot{m}(H_1 - H_2) \tag{6.15}$$

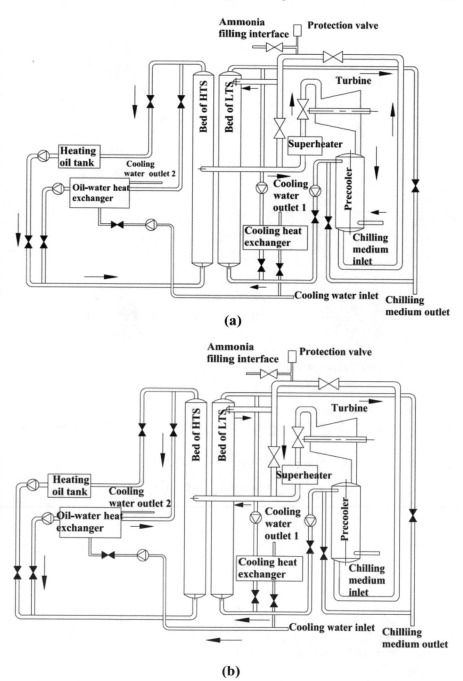

Fig. 6.20 The system design. **a** Electrical generation and sorption refrigeration process; **b** Resorption refrigeration process [40]

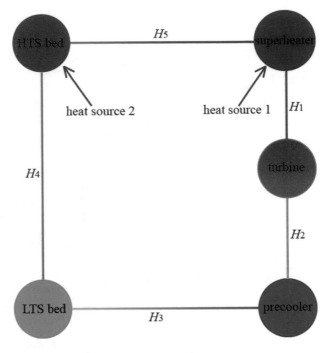

Fig. 6.21 Schematic diagram of the system [40]

where W is the electrical power generated by turbine (kW), $\dot{m}$ is the flow rate of the ammonia vapour (kg/s), H is the enthalpy of ammonia at different positions (kJ/kg).

$$Q_{de} + Q_h = \left(\Delta H_{deH} + \dot{m}\,H_5\right) + \dot{m}(H_1 - H_5)$$
$$= \left(\dot{m}\,\Delta H_{HTS} + \frac{m_{ad}C_p \times (T_{deH} - T_{hre})}{t_H}\right) + \dot{m}(H_1 - H_5),$$

in which

$$T_{hre} = \frac{T_{deH} + T_{adH}}{2} \tag{6.16}$$

where Q_h is the heat provided by the heat source 1 (kW), Q_{de} is the heat from the heat source 2 (kW). Q_{de} includes two parts, one part is for the sensible heat for HTS, which is heated from the temperature after heat recovery (T_{hre}, K) to the temperature of desorption (T_{deH}, K). Another part is for the desorption enthalpy of HTS, i.e. ΔH_{HTS} (kJ/kg). ΔH_{deH} is the difference between the final enthalpy and beginning enthalpy of HTS and NH_3 compounds, m_{ad} is the mass of HTS (kg), T_{adH} is the temperature after sorption (°C), C_p is the specific heat of HTS (kJ/(kg · K)), t_H is the heating time (s).

$$\eta_{El} = \frac{W}{Q_{de} + Q_h} \tag{6.17}$$

where η_{El} is the theoretical energy coefficient of performance for electrical generation.

Since the temperature of the LTS bed is almost equal to the ambient temperature T_c, the corresponding equilibrium pressure in the LTS bed and the pre-cooler is lower. For this value, the expanded ammonia vapour depends on the superheat. Therefore, due to the enthalpy difference of the ammonia vapour, it can provide cooling capacity (Q_{ref1}, kW). The thermal balance equation is:

$$Q_{ref1} = \dot{m}(H_2 - H_3) \tag{6.18}$$

The resorption refrigeration stage is much simpler. First, the cooling source is used to cool the HTS bed, and its temperature is controlled at the T_{adH} determined by the resorption pressure. In this process, it sorbs ammonia vapour from the LTS bed. Under the pressure drop between the HTS and LTS, the LTS bed desorbs, and the desorption heat of the LTS generates cooling power. The cooling power equation is:

$$Q_{ref2} = \dot{m}(\Delta H_{deL} + H_4) = \dot{m}\,\Delta H_{LTS} \tag{6.19}$$

where Q_{ref2} is the refrigeration capacity (kW), ΔH_{LTS} is the enthalpy for desorption of LTS and ammonia compounds, and ΔH_{deL} is the difference between final enthalpy and beginning enthalpy of LTS-ammonia compounds (kJ/kg). The refrigeration coefficient of performance (COP_{ref}) is:

$$COP_{ref} = \frac{Q_{ref1} + Q_{ref2}}{Q_{de} + Q_h} \tag{6.20}$$

6.3.1.2 Performance Analysis for the Application in a Combined System

For the application of sorption system, its working performance is mainly affected by heat and mass transfer performance. The biggest resistance is the mass transfer performance of the resorption refrigeration process. Considering that $BaCl_2$ has better equilibrium resorption refrigeration performance than $PbCl_2$ and it has a higher resorption pressure than $CaCl_2$, it has a potential for good mass transfer performance [42], so $BaCl_2$ is selected as the halide in the LTS bed.

The bed of the resorption system is a shell-and-tube heat exchanger, as shown in Fig. 6.22. The system uses a consolidated sorbent of halide and expanded natural graphite to fill the heat transfer tube. The baffle plates are utilized to increase the flow rate and heat transfer coefficient of the heat exchange medium. The COP will be affected by the metal mass of the bed. The heating heat of the non-heat recovery cycle is calculated by the following equation

$$
\begin{aligned}
Q_{de} + Q_h &= \left(\dot{m} \, \Delta H_{HTS} + \frac{m_{ad} C_p \times (T_{deH} - T_{adH})}{t_H} + \frac{m_{me} C_{pme} \times (T_{deH} - T_{adH})}{t_H} \right) + \dot{m}(H_{am2} - H_{am1}) \\
&= \dot{m} \left(\Delta H_{HTS} + \frac{m_{ad} C_p \times (T_{deH} - T_{adH})}{\dot{m} \, t_H} + \frac{m_{me} C_{pme} \times (T_{deH} - T_{adH})}{\dot{m} \, t_H} + (H_{am2} - H_{am1}) \right)
\end{aligned}
$$

$$(6.21)$$

where m_{me} is the mass of metal, C_{pme} is the specific heat of metal.

The designed refrigeration power is 15 kW, and the system design maximum pressure is 3.5 MPa, which is 0.5 MPa higher than the working pressure of 3 MPa. Assuming that the cycle time is 30 min, the mass of different HTS is calculated which is generally less than 30 kg. The material of the bed is usually steel, because ammonia and copper are corrosive and some sorbents such as $NiCl_2$ and aluminium are corrosive. Assuming that the mass ratio of the sorbent is 80% and the density of the consolidated sorbent is 400 kg/m^3, when the mass of the sorbent is 30 kg, the metal mass of the HTS bed is 431.9 kg. For such a system, the mass of ammonia in the desorption time is about 13 kg, which is equivalent to the ammonia in the resorption stage, which can be calculated from the refrigeration power, cycle time and reaction heat of $BaCl_2$. The performance of the system without heat recovery process is calculated, and the results are shown in Fig. 6.23. The η_{El} (Fig. 6.23a) and COP_{ref} (Fig. 6.23b) are affected by the heat capacity of metal materials. η_{El} and COP_{ref} are between 0.072–0.126 and 0.33–0.53, respectively, which are 30%-38% lower than the data without considering the heat capacity of the metal. The performance can be improved by the heat recovery between the two high-temperature superconducting beds, and the equation is:

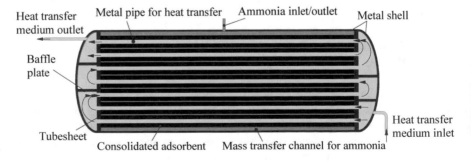

Heat transfer medium outlet   Metal pipe for heat transfer   Ammonia inlet/outlet   Metal shell

Baffle plate

Heat transfer medium inlet

Tubesheet   Consolidated adsorbent   Mass transfer channel for ammonia

Fig. 6.22 The tube-shell bed for sorption system [40]

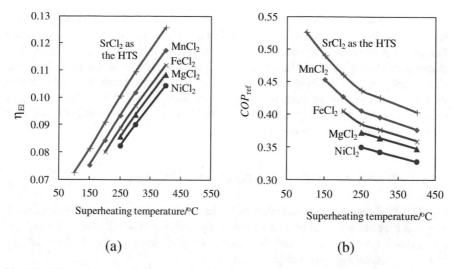

Fig. 6.23 The performance for the system without heat recovery. **a** η_{El} vs. superheating temperature; **b** COP_{ref} vs. superheating temperature [40]

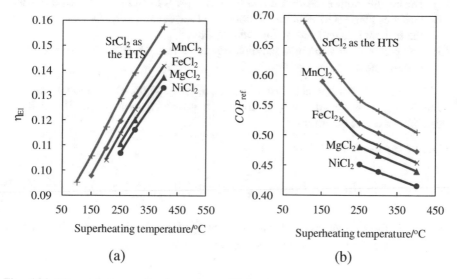

Fig. 6.24 The performance for the system with heat recovery, **a** η_{El} versus superheating temperature, **b** COP_{ref} versus superheating temperature [40]

$$Q_{de} + Q_h = \dot{m}\left(\Delta H_{HTS} + \frac{m_{ad}C_p \times (T_{deH} - T_{hre})}{\dot{m}\, t_H} + \frac{m_{me}C_{pme} \times (T_{deH} - T_{hre})}{\dot{m}\, t_H} + (H_1 - H_5)\right)$$

$$(6.22)$$

The performance of the heat recovery cycle is shown in Fig. 6.24, which takes into account the heat capacity of the metal and sorbent. Figure 6.24 shows that the ranges of η_{El} and COP_{ref} are 0.095–0.158 (Fig. 6.24a) and 0.416–0.691 (Fig. 6.24b), respectively. Compared with the result of the cycle without heat recovery, this value is significantly improved. Compared with the result of the cycle that does not consider the metal quality of the sorber, the value is reduced by about 15%-18%. The exergy efficiency of the system is calculated, and the highest value is about 0.82, which is 50% higher than the Goswami cycle.

6.3.1.3   Performance Calculation for Combined System

Figure 6.25 shows the schematic design of the unit tube of sorption bed in Fig. 6.22. The unit tube is composed of a unit tube, a layer of consolidated composite sorbent and a flowing heat transfer fluid. In order to simplify the mathematical model, the following assumptions need to be made:

(1) Heat and mass transfer only happened in the radial direction;
(2) There is no temperature difference between sorbent and refrigerant;
(3) During the sorption/desorption progress, the pressure of sorption bed is constant;
(4) Gaseous phase state equation follows the R-K equation and vapour velocity in the sorbent is calculated by the Darcy's law;
(5) Temperature variation in radial direction is neglected both for heat transfer fluid and the tube.

The mass transfer balance differential equations and one-dimensional unsteady state energy balance are used as the main governing equations. The detailed descriptions are as the follows:

(1) Energy balance of the sorption bed.

$$\varepsilon_a \frac{\partial}{\partial t}(\rho_g c_{p,g} T_g) + (1 - \varepsilon_a)\rho_s c_{p,s}\frac{\partial T_s}{\partial t} = \frac{1}{r}\frac{\partial}{\partial r}(r\lambda_{eff,s}\frac{\partial T_s}{\partial r}) + (1 - \varepsilon_a)\rho_a \Delta H \frac{\partial x}{\partial t} \quad (6.23)$$

where x is the sorption rate, r is radius of single tube, λ_{eff} is the effective thermal conductivity of sorbent, ε_a is the porosity of sorbent.

Fig. 6.25 Schematic design of the unit tube sorption bed [40]

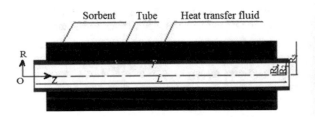

$$\frac{dN_g}{dt} = 4\pi r_i^2 K_{St}(\frac{P_i}{P_{Eq_{st}}} - 1)^{Ma}; x = 1 - (\frac{r_c}{r_g})^3 \tag{6.24}$$

where N_g is molecular sorption rate (mol/mol), K_{st} is dynamic coefficient, r_g is radius of grain, r_c is the radius of reaction area, ε_a is the porosity of sorbent, P_i is gas pressure correspond to reaction area.

(2)   Mass balance of the sorption bed.

$$\varepsilon_a \frac{\partial \rho_g}{\partial t} = -\frac{1}{r}\frac{\partial}{\partial r}(r\rho_g v_{g,r}) - (1 - \varepsilon_a)\rho_a \frac{\partial x}{\partial t} \tag{6.25}$$

where $v_{g,r}$ is the radial vapour diffusive velocity, which is determined by the equation of gas motion and the Darcy law.

$$v_{g,r} = -\frac{k}{\mu_g}\frac{\partial P_g}{\partial r} \tag{6.26}$$

The Gaseous phase state equation follows the R-K equation, where the error is under 0.8% for the NH$_3$ gas in such sorption system.

$$\rho_g = \frac{P_g}{ZRT_g} \tag{6.27}$$

where Z is the compressibility factor, which is calculated by P_{r0} and T_{r0}.

(3)   Energy balance of the heat transfer fluid and metal tube.

$$\frac{\partial T_f}{\partial t} + \frac{h_{fm}A_{fm}}{\rho_f c_{p,f} V_f}(T_f - T_m) = 0 \tag{6.28}$$

$$\frac{\partial T_m}{\partial t} + \frac{h_{fm}A_{fm}}{\rho_m c_{p,m} V_m}(T_m - T_f) + \frac{h_{ms}A_{ms}}{\rho_m c_{p,m} V_m}(T_m - T_{ms}) = 0 \tag{6.29}$$

(4)   The initial and boundary conditions.

Initial conditions:

$$T_s|_{\tau=0} = T_m|_{\tau=0} = T_f|_{\tau=0} = T_{amb}; P_s|_{\tau=0} = P_{amb} \tag{6.30}$$

Boundary conditions:

$$-\lambda_{eff}\frac{\partial T_s}{\partial r}|_{r=Ro} = h_{ms}(T_m - T_s); \frac{\partial T_s}{\partial r}|_{r=Ra} = 0 \tag{6.31}$$

$$\frac{\partial P_s}{\partial r}\Big|_{r=Ro} = 0; \quad P_s|_{r=Ra} = P_{cond}/P_{evp} \tag{6.32}$$

The system performances of such a unit tube sorption bed can be reflected with the coefficient of performances (COP) and specific cooling power (SCP), and they are calculated as follows:

$$COP = \frac{Q_e}{Q_h + Q_{de}} \tag{6.33}$$

$$SCP = \frac{Q_e}{m_s \tau_{cycle}} \tag{6.34}$$

where Q_h and Q_{de} are the input heat of system, Q_e is the refrigeration capacity, τ_{cycle} is the time of one cycle.

The above-mentioned one-dimensional partial differential equation is solved by the fully implicit difference method.

Table 6.4 lists the main related data used for the calculation of single tube. For each compound sorbent, thermal property is chosen by density 400 kg/m³ and ratio 75%. Table 6.5 shows the results of SCP and COP of 5 kinds of sorbents system. BaCl₂ compound sorbents gain the highest SCP and COP values. For different sorbents, SCP and COP range from 74.9–79.4 W/kg, 0.16–0.21, respectively.

As the single tube is calculated, four working pairs are also chosen to do the performance calculation of resorption. Table 6.6 show the results of SCP and COP of the four working pairs. FeCl₂-BaCl₂ shows the best performance with SCP of

Table 6.4 Parameters for calculation [40]

Parameter	Symbol	Value
Tube internal radius	R_i	10 mm
Tube external radius	R_o	12 mm
Sorbent bed external radius	R_a	24 mm
Sorbent bed axial length	L	1000 mm
Heat transfer fluid velocity	u_f	0.16 m/s
Compound sorbent packing density	ρ_s	577.1 kg/(m · K³)
Metal/ compound sorbent heat transfer coefficient	h_{ms}	300 W/(m² · K)
Porosity	ε	0.3
NH₃ vapour dynamic viscosity	μ	1.016×10^{-5} Pa·s
Cycle maximum temperature	T_{max}	200 °C
Cycle minimum temperature	T_{min}	30 °C
Evaporating temperature	Te	5 °C/–15 °C
Condenser temperature	T_{con}	35 °C/30 °C
Inlet heat transfer fluid temperature during heating	$T_{in, h}$	200 °C

Table 6.5 System performance of the 5 different kinds of sorbents [40]

Salts	Air conditioning condition			Freezing condition		
	Cycle time (min)	SCP (W/kg)	COP	Cycle time (min)	SCP (W/kg)	COP
$FeCl_2$	9.71	93.9945	0.4822	5.47	74.9095	0.3215
$MnCl_2$	9.89	95.5342	0.502	6.70	76.2946	0.3485
$SrCl_2$	10.05	96.4423	0.5103	7.88	77.32	0.3696
$NiCl_2$	10.38	97.2351	0.5411	8.96	78.15	0.4023
$BaCl_2$	10.72	98.6574	0.5802	9.47	79.4085	0.4338

Table 6.6 System performance of different working pairs [40]

Working pairs	Air conditioning condition			
	Cycle time (min)	SCP (W/kg)	COP	Power generation/kW
$MnCl_2$-$BaCl_2$	10.71	40.15	0.22	0.34
$SrCl_2$-$BaCl_2$	12.19	46.01	0.25	0.40
$NiCl_2$-$BaCl_2$	15.35	50.42	0.25	0.42
$FeCl_2$-$BaCl_2$	18.28	54.89	0.27	0.45

54.89 W/kg, COP of 0.271 and power generation of 0.45 kW. For different working pairs, SCP and COP range from 40.15–54.89 W/kg, 0.22–0.27, and 0.34–0.45 kW, respectively.

6.3.2 An Optimized Chemisorption Cycle for Power Generation

6.3.2.1 Thermodynamic Principle of Resorption Power Generation Cycle

A resorption cycle uses two different sorbents to combine the sorbent pairs [43, 44]. The basic resorption power generation (RPG) cycle is shown in Fig. 6.26. Its typical structure is a single-effect resorption cycle with an expander between the two sorption reactors. The ammonia vapor desorbed from one reactor under high temperature and high pressure enters to the expander, generates mechanical power in the expander, and is finally sorbed by another reactor. As shown in Fig. 6.26a and b, the RPG cycle can achieve continuous recovery of the waste heat of the power output in two half cycles, similar to the quasi-continuous output. If synthesis occurs between the expanded exhaust and the HTS under high pressure, a large amount of upgrade heat may also be generated in the second half of the cycle.

The thermodynamic processes of the RPG cycle are plotted in the Clapeyron diagram in Fig. 6.27. In the first half cycle, the HTS reactor is heated (point A in

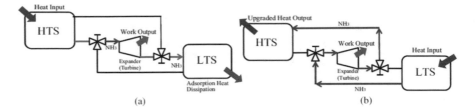

Fig. 6.26 Basic RPG cycle. **a** First half-cycle: heat input to HTS and power generation; **b** Second half-cycle: heat input to LTS, power generation and potentially heat upgrade [41]

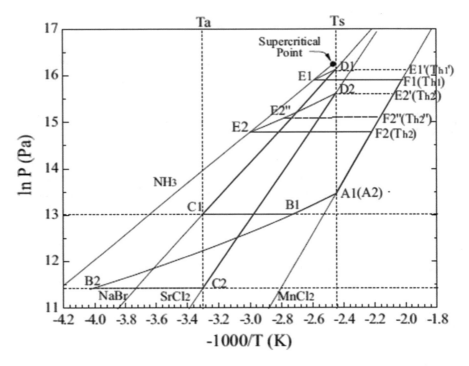

Fig. 6.27 Clapeyron diagram of basic RPG cycles, using MnCl₂-NaBr and MnCl₂-SrCl₂ sorbent pairs as examples [41]

Fig. 6.27) to desorb ammonia at high temperature and high pressure. Due to pressure and temperature, ammonia has the potential to expand through the expander (process A–B, isentropic expansion) with work output. And then at the heat sink temperature T_a (point C), the exhausted ammonia gas is sorbed by LTS. Therefore, the vapor expansion is affected by the LTS sorption pressure of T_a as the back pressure. In the second half cycle, the LTS reactor uses the same heat source to perform heating (C–D) and desorption while HTS reactor is cooled at the heat sink temperature. Since the equilibrium pressure of LTS is higher than that of HTS at the same temperature, the pressure (point D) of the ammonia released by the LTS reactor is much higher than that in the first half of the cycle, and the vapor passes through the expander for the second power generation. The backpressure of this second expansion is HTS sorption pressure with the heat sink temperature, which is a suitable condition for vapor expansion. On the other hand, the ammonia expansion is limited by the condition of durable expander operation, thus the expansion will terminate when the working fluid reaches saturation state as shown in Fig. 6.27. Finally, the sum of processes A-B and D–E represents the total potential work output of a complete RPG cycle. Since the pressure ratio of the $MnCl_2$-$SrCl_2$ RPG cycle is greater, in the first half of the cycle, the RPG cycle of the $MnCl_2$-$SrCl_2$ pair has a greater power output than the $MnCl_2$-$NaBr$ pair (A1B1 < A2B2 in Fig. 6.27). In the second half cycle, although the pressure ratio of the $MnCl_2$-$NaBr$ RPG cycle is greater than that of the $MnCl_2$-$SrCl_2$ pair, the ammonia saturation conditions severely hinder the work output of the $MnCl_2$-$NaBr$ RPG cycle. Finally, under the same heat sink and heat source conditions, the performance of the $MnCl_2$-$SrCl_2$ pair is better than the $MnCl_2$-$NaBr$ pair (A1B1 + D1E1 < A2B2 + D2E2).

As the vapor expansion occurs at the high pressure of the second half cycle (point E), the exothermic sorption between the HTS bed and the discharged ammonia occurs at a high equilibrium temperature (point F), which realizes heat transformation. If the heat upgrade is desired as much as possible, the released ammonia can be directly sorbed by the HTS reactor at a higher pressure (at point E') without the need for an expansion process. Of course, the energy output (heat or power) can be balanced by adjusting the expansion rate according to specific needs. For example, the slightly modified D2-E2-F2 cycle in the dotted line in Fig. 6.27 reduces some of the work output of the larger thermal energy upgrade.

6.3.2.2 Thermodynamic Principle of Advanced RPG Cycle

In order to maximize the effective pressure ratio of the RPG cycle, a reheating process is proposed. This reheating method describes the process of additionally heating the desorbed ammonia vapor before it expands. Because the method is performed under the premise of the optimal desorption temperature, the reheating conditions are complicated.

Figure 6.28 shows the increase in power generation based on the desorption-expansion with reheating process when NaBr ammoniate is coupled with $SrCl_2$. As the state of the expanded exhaust gas should fall in the gray-marked area in Fig. 6.28a, the two main limiting conditions for expansion in the RPG cycle are drawn in the Clapeyron diagram, which are backpressure and ammonia saturation condition. Due to the univariate characteristic of the chemisorption cycle, the desorption temperature/pressure can be adjusted before isobaric reheating to obtain the maximum work output. The four different desorption processes in Fig. 6.28a use temperatures of 110 °C, 80 °C, 40 °C, and 25 °C, respectively. The ammonia vapor after desorption is then subsequently reheated isobaricly to a heat source temperature of 110 °C. Obviously, the isentropic work output first increases as the desorption temperature decreases (1–2 < 2'-3' < 2'-3''), because of the increasing deviation of the state of

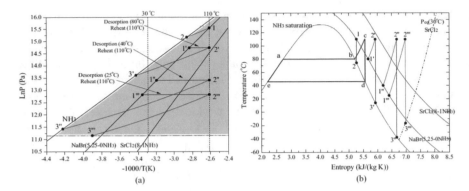

Fig. 6.28 Reheating method to improve the power generation in advanced RPG cycle. **a** Clapeyron P-T diagram; **b** T-S diagram [41]

the expander inlet ammonia from the ammonia saturation line. Due to the limitation of the sorption equilibrium pressure of $SrCl_2$, as the desorption temperature further decreases, the output work gradually decreases. Therefore, it can be deduced that at a given heat source temperature, there is an optimal desorption temperature point for maximum work output. The same conclusion can be obtained from the T-S diagram shown in Fig. 6.28b.

Reasonable management of heat energy utilization at different temperature can effectively realize the reheating process. A common situation in industrial processes is that there are several streams of waste heat available at different temperatures. As shown in Fig. 6.29a, in this case, the waste heat for the reheating and desorption process can be arranged separately. When there is only a single heat source at a certain temperature, the heat source should serve the reheating heat exchanger before the desorption reactor, as shown in Fig. 6.29b. Meanwhile, the flow rate of the heat transfer fluid and heat exchange area should be carefully determined to achieve the designed condition. Alternatively, if the optimal desorption temperature is lower than the ambient temperature, cooling energy can be generated by desorption at a lower temperature, as shown in Fig. 6.29c.

6.3.2.3   Analytical Methodology

The cycle performance with heat sink temperature at 25 °C and the heat source between 40 and 200 °C is evaluated via the following equations with chemical and physical parameters of sorbent listed in Table 6.7. The resorption working pairs are grouped between three typical sorbents, which can be formulated as Eqs. 6.35–6.37.

$$MnCl_2 \cdot 2NH_3 + 4NH_3 \Leftrightarrow MnCl_2 \cdot 6NH_3 + 4\Delta H_r \qquad (6.35)$$

$$SrCl_2 \cdot NH_3 + 7NH_3 \Leftrightarrow SrCl_2 \cdot 8NH_3 + 7\Delta H_r \qquad (6.36)$$

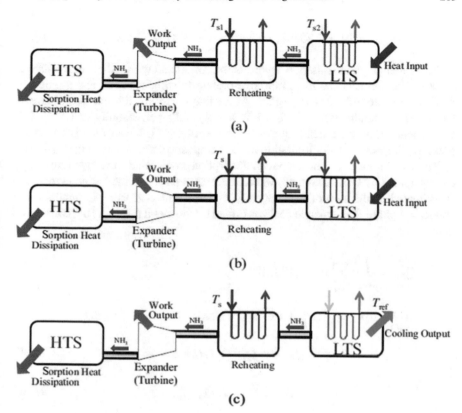

Fig. 6.29 Different arrangements of advanced RPG cycle. **a** Two heat sources at different temperatures; **b** Single heat source; **c** Single heat source, LTS desorption at low temperature with cooling output [41]

Table 6.7 Thermodynamic parameters of ammonia and different sorbents [47]

	ΔH_r J/mol	ΔS_r J/(mol·K)	Max. uptake (g/g)	c_p (J/(kg·K))
NH_3	23,366	150.52	–	–
$MnCl_2$ (2–6 NH_3)	47,416	228.07	0.540	$\frac{4.184}{M_{MnCl_2}} \cdot (16.2 + 0.0052 \cdot T\,(K))$
NaBr (0–5.25 NH_3)	30,491	208.8	0.867	$\frac{4.184}{M_{NaBr}} \cdot (11.74 + 0.00233 \cdot T\,(K))$
$SrCl_2$ (1–8 NH_3)	41,431	228.8	0.751	$\frac{4.184}{M_{SrCl_2}} \cdot (18.2 + 0.00244 \cdot T\,(K))$

$$NaBr + 5.25NH_3 \Leftrightarrow NaBr \cdot 5.25NH_3 + 5.25\Delta H_r \qquad (6.37)$$

Expanded graphite is used as a porous matrix, mixed with halide ammoniate as a composite sorbent. The mass ratio of graphite to sorbent is 1:3, and the density of the composite sorbent is 450 kg/m^3. According to reports, this composite sorbent has a thermal conductivity of 1.7–3.1 W/(m·K) and a permeability of 10^{-12}–10^{-11} m^2, depending on the amount of ammonia sorption [45]. Under the high working pressure, the mass transfer limitation in the composite sorbent can be ignored [46].

The sensible heat, total heat input for desorption process and corresponding exergy are calculated by Eqs. 6.38–6.40, where the heat capacity of the halide ammoniate is estimated as the sum of heat capacity of the halide and that of ammonia in a condensed phase. The sensible heat consumed by the metal reactor is not considered here.

$$Q_{sen} = \int_{T_1}^{T_2} \sum_{i=1}^{3} (m \cdot c_p(T))_i dT$$
$$= \int_{T_1}^{T_2} \left[(m \cdot c_p(T))_{EG} + (m \cdot c_p(T))_{halide} + (m \cdot c_p(T))_{NH_3} \right] dT \qquad (6.38)$$

$$Q_{input} = Q_{sen} + \Delta H_r \cdot \Delta x \qquad (6.39)$$

$$E_{input} = (1 - T_a/T_s) Q_{input} \qquad (6.40)$$

The consumed energy and the corresponding exergy in the desorption process are calculated by Eqs. 6.41–6.44:

$$Q_{sen,opt} = \int_{T_1}^{T_{opt}} \sum_{i=1}^{3} (m \cdot c_p(T))_i dT$$
$$= \int_{T_1}^{T_{opt}} \left[(m \cdot c_p(T))_{EG} + (m \cdot c_p(T))_{halide} + (m \cdot c_p(T))_{NH_3} \right] dT \qquad (6.41)$$

$$Q_{reheat} = m_{NH_3} \cdot (h(T_s) - h(T_{opt})) \qquad (6.42)$$

$$Q_{input} = Q_{sen, opt} + \Delta H_r \cdot \Delta x + Q_{reheat} \qquad (6.43)$$

$$E_{input} = (1 - T_a/T_{opt}) \cdot (Q_{sen,opt} + \Delta H_r \cdot \Delta x) + (1 - T_a/T_s) \cdot Q_{reheat} \qquad (6.44)$$

When the optimal desorption temperature is lower than the ambient temperature as shown in Fig. 6.29c, the potential cooling production can be estimated by Eqs. 6.45 and 6.46.

$$Q_{\text{ref}} = \Delta H_r \cdot \Delta x - \left[(m \cdot c_p(T))_{\text{EG}} + (m \cdot c_p(T))_{\text{halide}} + (m \cdot c_p(T))_{\text{NH}_3} \right] \cdot (T_a - T_{\text{opt}}) \quad (6.45)$$

$$E_{\text{ref}} = (T_a / T_{\text{opt}} - 1) Q_{\text{ref}} \quad (6.46)$$

The mechanical energy output from the assumed isentropic expansion can be expressed as:

$$W = E_w = m_{\text{NH}_3} \cdot (h_{\text{out}} - h_{\text{in}}) \quad (6.47)$$

The equations of energy efficiency and exergy efficiency are shown as follows, where the subscript 1 and 2 denote the first half-cycle and the second half-cycle.

$$\eta_{\text{en,w}} = (W_1 + W_2)/(Q_{\text{input1}} + Q_{\text{input2}}) \quad (6.48)$$

$$COP = Q_{\text{ref}}/(Q_{\text{input1}} + Q_{\text{input2}}) \quad (6.49)$$

$$\eta_{\text{ex,w}} = (E_{\text{w1}} + E_{\text{w2}})/(E_{\text{input1}} + E_{\text{input2}}) \quad (6.50)$$

$$\eta_{\text{ex,ref}} = E_{\text{ref}}/(E_{\text{input1}} + E_{\text{input2}}) \quad (6.51)$$

It is arguable that the total energy efficiency of refrigeration output and work output expressed in Eq. 6.52 overestimates the efficiency of the cycle because the capacity of the refrigeration output is not considered. In order to compare different combined power and cooling cycles more fairly, according to the suggestion of literature [48], the energy efficiency is calculated by Eq. 6.53. The exergy efficiency of the combined cycle can be calculated by Eq. 6.54.

$$\eta_{\text{en}} = \eta_{\text{en,w}} + COP \quad (6.52)$$

$$\eta_{\text{en}} = \eta_{\text{en,w}} + E_{\text{ref}}/(Q_{\text{input1}} + Q_{\text{input2}}) \quad (6.53)$$

$$\eta_{\text{ex}} = \eta_{\text{ex,w}} + \eta_{\text{ex,ref}}. \quad (6.54)$$

6.3.2.4 The Optimal Desorption Temperature and Work Output

Figure 6.30 shows the work output of 1.0 kg reacted ammonia. Figure 6.30a, c and e show the work output of various sorbents used in the first half of the RPG cycle, and Fig. 6.30b, d and f show the work output of the second half of the cycle. According to the mono-variant characteristics and thermodynamic balance of HTS and LTS ammoniates, a higher heat source temperature is required in the first half of the

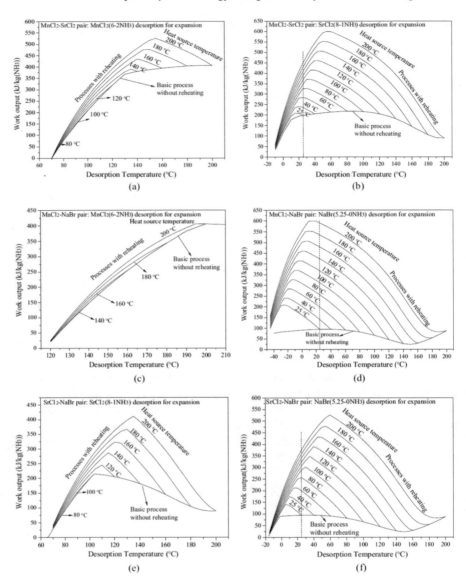

Fig. 6.30 Work outputs of RPG cycles with different sorbent pairs under conditions of different desorption temperature and reheating temperature [41]

cycle to drive the desorption of HTS ammoniates and vapor expansion as well as provide enough sorption pressure for LTS ammoniates. For the MnCl₂-NaBr pair, the threshold of such a desorption temperature for effective work output is at around 120 °C, and for the other two working pairs is about 70 °C. In contrast, due to the higher desorption pressure of LTSs and the lower sorption pressure of HTSs, the desorption temperature of the second half of the cycle can be lower than zero, which indicates that the pressure difference conducive to power generation is much larger than that of the first half of the cycle.

The single red curve at the bottom of each graph labeled "Basic Process Without Reheating" represents the original basic RPG cycle. A series of black curves imply the potential improvements through the introduction of the reheating process. Each black curve represents the changing work output when ammonia is desorbed at a lower temperature and then reheated to a given heat source temperature. As the desorption temperature increases, the sorption work also increases. It is because the pressure difference between desorption and sorption increases. Then, when the heat source temperature is high enough, as the desorption temperature continues to rise, the work-output curve reaches its peak and then shows a downward trend. In this case, when the apex is lower than a given heat source temperature, the optimal desorption temperature for maximum work output appears. Otherwise, take the given heat source temperature as the best desorption temperature. The main reason for the decline of the work output curve is the limitation of ammonia saturation conditions. $SrCl_2$-NH_3 and $NaBr$-NH_3 reaction equilibrium lines tend to converge with the increase of temperature, that is, the state of desorbed ammonia is getting closer and closer to the saturated state, resulting in little space for dry vapor expansion and therefore the work output is reduced. It can be seen from Fig. 6.30 that the best desorption temperature for HTS is mainly above 100 °C, while for LTS desorption, the desorption temperature is mostly not higher than 60 °C. It is worth noting that in the second half of the cycle, as the desorption temperature continues to rise, the work output will reach their respective troughs, but then tends to rebound. This is because the desorbed ammonia enters the supercritical region at some points, which means that the expansion potential beyond saturation conditions is greater.

Since the expansion of the ammonia desorbed by LTS is more restricted by saturation conditions than the ammonia desorbed by HTS, the implementation of reheating can make the second half of the cycle have a more obvious improvement than the first half of the cycle, as shown in Fig. 6.30. For example, when the heat source temperature is 200 °C, the maximum output power in the first half cycle can be increased by 2–200% according to different working pairs, while the maximum work output in the second half cycle can be increased by 500–600%.

(1) Basic RPG cycle

Figure 6.31 shows the pressure ratio of the basic RPG cycle using different working pairs. The pressure ratio of the first half cycle is much lower than the pressure ratio of the second half cycle. When the desorption temperature reaches about 100 °C, these values all soar to the magnitude of hundreds in the second half of the cycle. However, the actual expansion rate of the RPG cycle shown in Fig. 6.32 is much lower. It should be noted the difference between "pressure ratio" and "expansion ratio". The former represents external conditions that facilitate or restrict the vapor expansion, and the latter reflects the actual thermodynamic state evolution of the working fluid for steam expansion.

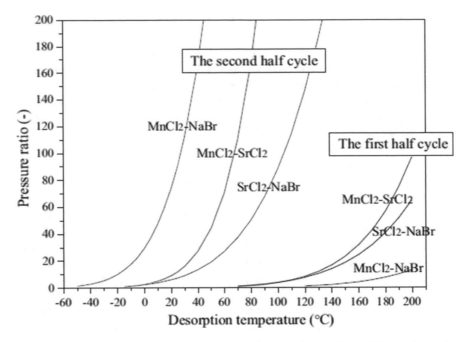

Fig. 6.31 Pressure ratio of RPG cycles using different sorbent pairs at different desorption temperature [41]

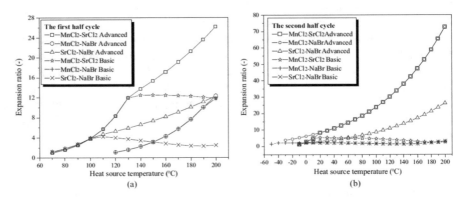

Fig. 6.32 Expansion ratio of the basic RPG cycles and the advanced RPG cycles **a** in the first half-cycle; **b** in the second half-cycle [41]

Generally, a larger pressure ratio is beneficial to productive generation through vapor expansion, as long as dry expansion can be guaranteed and the expander can handle this expansion ratio. Although the pressure ratio of the first half cycle is much smaller than the pressure ratio of the second half cycle, the expansion ratio of the first half cycle is greater than the expansion ratio of the second half cycle, making the power generation in the first half cycle relatively high. The key reason for this opposite phenomenon is the limitation of ammonia saturation conditions. The curve of the ammonia vapor desorbed from LTS ammoniate is closer to the ammonia saturation curve. Therefore, even if external conditions are conducive to vapor expansion, the nature of the wet fluid limits the expansion, causing most of the power generation potential to be wasted. The ammonia desorbed from HTS ammoniate is relatively far from the ammonia saturation curve. Although the pressure ratio is relatively small, there is still more room for vapor expansion.

(2) Advanced RPG cycle with reheating

In Fig. 6.32a, only at the high heat source temperature of the first half cycle the expansion rate can be improved by reheating. The main constraint is the backpressure at low heat source temperature, which then becomes saturated under high temperature heat source. As shown in Fig. 6.32b, in the second half of the cycle, the improvement of the expansion rate is more obvious, from about 5 to dozens. In many cases, the optimal expansion process wants to end at the intersection of the ammonia saturation line and the backpressure line, as the end point of the bottom left hand side of the grey region shown in Fig. 6.28. Therefore, there is overlap between some of the curves in Fig. 6.32. These curves use the same halide ammoniate downstream of the expander, so the ammonia saturation line and the backpressure line are both fixed with the same heat source.

As shown in Fig. 6.30d and f, some of the optimal desorption temperatures in the advanced RPG cycle are much lower than the ambient temperature, especially for the $MnCl_2$-NaBr pair. When the heat source temperature is between 25 and 200 °C, the best desorption temperature is -28–9 °C. The results show that a considerable cooling product is obtained by LTS desorption, and the maximum work output is obtained after the reheating process. In other words, with one desorption heat input in the first half-cycle and two small sensible heat input for reheating in each half-cycle, an advanced RPG cycle can obtain two work outputs and large amount of cooling energy. In addition, it is worth noting that the differences between the proposed advanced RPG and the resorption power and cooling cogeneration cycle reported in Reference [42] are as follows: (1) The current cycle obtains one more work output in the second half cycle. (2) A reheating process is utilized on the basis of the identification of the optimal desorption temperature to achieve maximum work output. The reheating process is suitable for the first half of the cycle and the second half of the cycle. It can continuously recover the waste heat and has a large energy utilization efficiency.

Figure 6.33 clearly shows the maximum potential and the significant advantages of the advanced RPG cycle compared to the basic RPG cycle under different operating conditions. Generally speaking, depending on the working pair and the heat source temperature, the reheating concept can increase the total working output during the entire operating cycle by 10–600%. When the heat source temperature is higher than the optimal desorption temperature, the work output increases linearly with the heat source temperature. Under the conditions studied, the specific power generation capacity of $MnCl_2$-$SrCl_2$ pair is better than the other two working pairs. When the

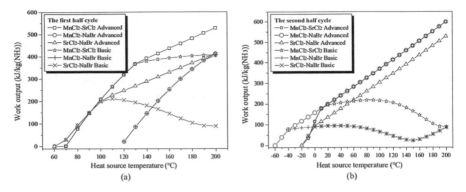

Fig. 6.33 Work output in basic RPG cycles and advanced RPG cycles, **a** the first half-cycle; **b** the second half-cycle [41]

heat source temperature is 200 °C, the first half cycle can produce 540 kJ/kg (NH_3), and the second cycle produces about 600 kJ/kg of NH_3.

6.3.2.5 Thermal Efficiency, Exergy Efficiency and Energy Density

Obviously, in an advanced RPG cycle, more work output will lead to higher system efficiency. In addition, in the advanced RPG cycle, the solid sorbent consumes less sensible heat, which can further improve the efficiency due to the lower desorption temperature. Therefore, for the calculation of the Eqs. 6.48 and 6.50, the collective effect of decreasing the denominator and increasing the numerator can greatly increase the final efficiency value. The sensible heat load of the metal reactor is not considered in the evaluation, otherwise the use of advanced cycles would increase the efficiency more significantly.

Figure 6.34 shows the exergy and thermal efficiency of power generation. The two half cycles use the same heat source to simulate the continuous recovery of given low-grade heat to achieve quasi-continuous work output. When the heat source temperature is between 70 and 200 °C, the thermal efficiency of work output based on the advanced RPG about 6–24%, which is about 41–63% of the Carnot cycle efficiency. As shown in Fig. 6.33, the work output of the $MnCl_2$-NaBr pair is not the highest, but the work output thermal efficiency of this working pair is the best, as shown in Fig. 6.34a. This is because the optimal desorption temperature of the second half cycle is lower than the ambient temperature, so this part of the desorption heat can be ignored. Depending on different work pairs, the exergy efficiency is between 50–85%. Compared with the basic RPG cycle, when the heat source temperature is 200 °C, the thermal efficiency of the advanced RPG cycle is increased by 1.4 times, 2.0 times and 4.6 times and the exergy efficiency is increased by 2.9 times, 2.0 times and 8.3 times, respectively for $MnCl_2$-$SrCl_2$, $MnCl_2$-NaBr, and $SrCl_2$-NaBr.

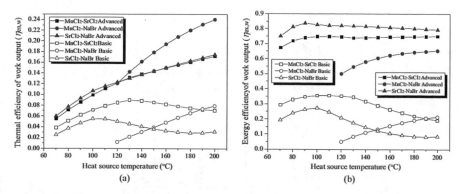

Fig. 6.34 Work output efficiencies of the basic RPG cycles and the advanced RPG cycles, **a** thermal efficiency; **b** exergy efficiency [41]

If desorption occurs at sub-ambient temperatures, advanced APG cycles can further reduce energy input and obtain additional cooling energy. Table 6.8 shows the efficiencies of advanced RPG cycles which have combined power with cooling output at optimal desorption temperatures. By adjusting the working conditions, the cooling capacity and the work output can be balanced according to the bespoken requirement, such as the work output could be compromised to reach a lower refrigeration temperature. The total energy efficiency (η_{en}) takes energy quality into account, as shown in Eq. 6.53. At the same time, the exergy efficiency value related to the additional cooling amount is used to evaluate the overall efficiency instead of the COP value.

Table 6.9 lists the mechanical energy density of RPG cycles using different working pairs, which is defined as the mechanical energy generated by one cycle per unit volume of sorbent. Since the research focus on the optimization of cycle thermodynamics and the exploration of the maximum potential to provide more generic information, specific reactor design and heat exchanger design are not considered in the density assessment. If a complete system prototype design with other key components is envisaged, the volume of the entire prototype for energy density calculation can provide more practical insights.

6.3.2.6 Comparison with Other Bottoming Cycles Driven by Low Grade Heat for Power Generation

With some of the optimal desorption temperatures being found to be lower than the ambient temperature, if the advanced RPG cycle is only used for power generation, the desorption of the second half of the cycle can be deliberately implemented at ambient temperature to minimize heat input and further improve energy efficiency. In addition, this method simplifies the configuration and operation of the system. The thermal efficiency of the advanced RPG cycle using this operating strategy is

Table 6.8 Performance of advanced RPG cycles for combined power and cooling output [41]

Sorbent pairs	Heat source temperature in first half-cycle (°C)	LTS desorption temperature (refrigeration temperature) (°C)	Reheating temperature (in second half-cycle)	Working generation efficiency ($\eta_{en,w}$)	COP (refrigeration)	Total energy efficiency (η_{en})	Total exergy efficiency (η_{ex})
MnCl$_2$-SrCl$_2$	180	0	80	0.169	0.623	0.226	0.791
	180	10.5	80	0.195	0.628	0.228	0.790
MnCl$_2$-NaBr	180	−20	120	0.173	0.420	0.547	0.740
	180	−10	120	0.188	0.451	0.248	0.741
	180	0	80	0.169	0.478	0.213	0.637
	180	10	80	0.165	0.490	0.191	0.570

Table 6.9 Mechanical energy density with different working pairs based on one complete cycle [41]		Basic RPG cycle (MJ/m^3)	Advanced RPG cycle (MJ/m^3)
	MnCl$_2$-SrCl$_2$	3.7–52.6	30.1–120.0
	MnCl$_2$-NaBr	4.4–55.3	24.7–113.7
	SrCl$_2$-NaBr	6.2–24.2	37.0–128.9

compared with other bottom cycles (including organic Rankine cycle (ORC), Stirling cycle, and thermoelectric cycle (TE)). It has been parametrically investigated by Bianchi et al. [49]. The analysis model of Bianchi et al. is based on the following reasonable assumptions:

(1) The temperature difference between the evaporation temperature of the working fluid and the available heat source temperature is fixed at 50 °C.
(2) The temperature of the heat sink is 15 °C, and the condensation temperature of the working fluid is 33 °C.
(3) The isentropic efficiency of the expander is 75%, and the mechanical efficiency is 97%.

Figure 6.35 compares these three cycles with the work of Bianchi et al., with the current advanced RPG cycle using the same analytical assumptions described above.

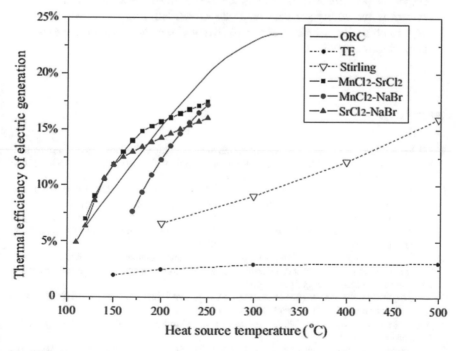

Fig. 6.35 Comparison of advanced RPG cycles to other bottoming cycles for power generation [49]

Under the various working conditions studied, the performance of the advanced RPG cycle is significantly better than the stirling cycle and the TE cycle. When the heat source temperature is between 120 and 200 °C, the advanced RPG cycle of $MnCl_2$-$SrCl_2$ and $SrCl_2$-NaBr have higher thermal efficiency than the ORC cycle.

Based on the changes in Fig. 6.35, and combined with theoretical principles, it is reasonable to speculate that:

(1) When the heat source temperature is higher than 250 °C, the thermal efficiency of $MnCl_2$-NaBr advanced RPG cycle is the highest among these three advanced RPG cycles;

(2) Driven by a low-temperature heat source, a working fluid pair composed of two closer equilibrium halide amides is more likely to obtain higher thermal efficiency, because such a resorption cycle can produce moderate pressure ratio between desorption and sorption in both in two half cycles.

(3) When the heat source temperature is high enough due to the larger pressure ratio produced in the second half cycle, the working pair composed of two halide ammoniates which have bigger equilibrium disparity between them tends to have higher thermal efficiency. Therefore, the proposed advanced RPG cycle can flexibly adopt different working pairs for different application scenarios. Hundreds of halides are worth exploring by using this resorption cycle for low-grade heat recovery, and their in-depth knowledge and more beneficial characteristics need further research. Another significant flexibility of the sorption system is the ability for heat upgrades or refrigeration or power generation or a combination, which is very attractive and has the potential to expand its applications broadly.

References

1. Michel B, Mazet N, Neveu P (2014) Experimental investigation of an innovative thermochemical process operating with a hydrate salt and moist air for thermal storage of solar energy: global performance. Appl Energy 129:177–186
2. Jacob R, Bruno F (2015) Review on shell materials used in the encapsulation of phase change materials for high temperature thermal energy storage. Renew Sustain Energy Rev 48:79–87
3. Dheep GR, Sreekumar A (2015) Influence of accelerated thermal charging and discharging cycles on thermo-physical properties of organic phase change materials for solar thermal energy storage applications. Energy Convers Manage 105:13–19
4. Li T, Wang R, Kiplagat JK, Kang YT (2013) Performance analysis of an integrated energy storage and energy upgrade thermochemical solid-gas sorption system for seasonal storage of solar thermal energy. Energy 50:454–467
5. Zhou D, Shire GSF, Tian Y (2014) Parametric analysis of influencing factors in phase change material wallboard (PCMW). Appl Energy 119:33–42
6. Yu N, Wang RZ, Wang LW (2013) Sorption thermal storage for solar energy. Prog Energy Combust Sci 39:489–514
7. N'Tsoukpoe KE, Liu H, Pierrès NL, Luo L (2009) A review on long-term sorption solar energy storage. Renew Sustain Energy Rev 13:2385–2396

8. Li G, Qian S, Lee H, Hwang Y, Radermacher R (2014) Experimental investigation of energy and exergy performance of short term adsorption heat storage for residential application. Energy 65:675–691

9. Li G (2016) Sensible heat thermal storage energy and exergy performance evaluations. Renew Sustain Energy Rev 53:897–923

10. Aydin D, Utlu Z, Kincay O (2015) Thermal performance analysis of a solar energy sourced latent heat storage. Renew Sustain Energy Rev 50:1213–1225

11. Chen C, Liu W, Wang H, Peng K (2015) Synthesis and performances of novel solid-solid phase change materials with hexahydroxy compounds for thermal energy storage. Appl Energy 152:198–206

12. Liu W, Chen G, Yan B, Zhou Z, Du H, Zuo J (2015) Hourly operation strategy of a CCHP system with GSHP and thermal energy storage (TES) under variable loads: a case study. Energy & Buildings 93:143–153

13. Alam TE, Dhau JS, Goswami DY, Stefanakos E (2015) Macroencapsulation and characterization of phase change materials for latent heat thermal energy storage systems. Appl Energy 154:92–101

14. Farid MM, Khudhair AM, Razack SAK, Al-Hallaj S (2004) A review on phase change energy storage: materials and applications. Energy Convers Manage 45:1597–1615

15. Cot-Gores J, Castell A, Cabeza LF (2012) Thermochemical energy storage and conversion: a-state-of-the-art review of the experimental research under practical conditions. Renew Sustain Energy Rev 16:5207–5224

16. Balasubramanian G, Ghommem M, Hajj MR, Wong WP, Tomlin JA, Puri IK (2010) Modeling of thermochemical energy storage by salt hydrates. Int J Heat Mass Transf 53:5700–5706

17. Johannes K, Kuznik F, Hubert JL, Durier F, Obrecht C (2015) Design and characterisation of a high powered energy dense zeolite thermal energy storage system for buildings. Appl Energy 159:80–86

18. Korhammer K, Druske MM, Fopah-Lele A, Rammelberg HU, Wegscheider N, Opel O et al (2016) Sorption and thermal characterization of composite materials based on chlorides for thermal energy storage. Appl Energy 162:1462–1472

19. Yan T, Wang RZ, Li TX, Wang LW, Fred IT (2015) A review of promising candidate reactions for chemical heat storage. Renew Sustain Energy Rev 43:13–31

20. Li TX, Wang RZ, Yan T (2015) Solid-gas thermochemical sorption thermal battery for solar cooling and heating energy storage and heat transformer. Energy 84:745–758

21. Haije WG, Veldhuis JBJ, Smeding SF, Grisel RJH (2007) Solid/vapour sorption heat transformer: design and performance. Appl Therm Eng 27:1371–1376

22. Li TX, Wu S, Yan T, Xu JX, Wang RZ (2016) A novel solid-gas thermochemical multilevel sorption thermal battery for cascaded solar thermal energy storage. Appl Energy 161:1–10

23. Goetz V, Spinner B, Lepinasse E (1997) A solid-gas thermochemical cooling system using BaCl$_2$ and NiCl$_2$. Energy 22:49–58

24. Wang LW, Bao HS, Wang RZ (2009) A comparison of the performances of adsorption and resorption refrigeration systems powered by the low grade heat. Renew Energy 34:2373–2379

25. Zhu FQ, Jiang L, Wang LW, Wang RZ (2016) Experimental investigation on a MnCl$_2$ CaCl$_2$ NH$_3$ resorption system for heat and refrigeration cogeneration. Appl Energy 181:29–37

26. Jiang L, Wang LW, Jin ZQ, Tian B, Wang RZ (2012) Permeability and thermal conductivity of compact adsorbent of salts for sorption refrigeration. J Heat Transfer 134:104503

27. Jiang L, Wang LW, Wang RZ (2014) Investigation on thermal conductive consolidated composite CaCl$_2$ for adsorption refrigeration. Int J Therm Sci 81:68–75

28. Wang LW, Metcalf SJ, Critoph RE, Thorpe R, Tamainot-Telto Z (2012) Development of thermal conductive consolidated activated carbon for adsorption refrigeration. Carbon 50:977–986

29. Bao H, Wang Y, Roskilly AP (2014) Modelling of a chemisorption refrigeration and power cogeneration system. Appl Energy 119:351–362

30. Bao H, Wang Y, Charalambous C, Lu Z, Wang L, Wang R et al (2014) Chemisorption cooling and electric power cogeneration system driven by low grade heat. Energy 72:590–598

31. Lemort V, Quoilin S, Cuevas C, Lebrun J (2009) Testing and modeling a scroll expander integrated into an organic rankine cycle. Appl Therm Eng 29:3094–3102
32. Qiu G, Liu H, Riffat S (2011) Expanders for micro-CHP systems with organic rankine cycle. Appl Therm Eng 31:3301–3307
33. Quoilin S, Lemort V, Lebrun J (2010) Experimental study and modeling of an organic rankine cycle using scroll expander. Appl Energy 87:1260–1268
34. Aoun B, Clodic DF (2008) Theoretical and experimental study of an oil-free scroll vapor expander, Purdue University
35. Wang W, Qu TF, Wang RZ (2002) Influence of degree of mass recovery and heat regeneration on adsorption refrigeration cycles. Energy Convers Manage 43:733–741
36. Lu ZS, Wang LW, Wang RZ (2012) Experimental analysis of an adsorption refrigerator with mass and heat-pipe heat recovery process. Energy Convers Manage 53:291–297
37. Ng KC, Wang X, Lim YS, Saha BB, Chakarborty A, Koyama S et al (2006) Experimental study on performance improvement of a four-bed adsorption chiller by using heat and mass recovery. Int J Heat Mass Transf 49:3343–3348
38. Srivastava NC, Eames IW (1998) A review of adsorbents and adsorbates in solid-vapour adsorption heat pump systems. Appl Therm Eng 18:707–714
39. Tamainot-Telto Z, Metcalf SJ, Critoph RE, Zhong Y, Thorpe R (2009) Carbon-ammonia pairs for adsorption refrigeration applications: ice making, air conditioning and heat pumping. Int J Refrig 32:1212–1229
40. Jiang L, Wang LW, Roskilly AP, Wang RZ (2013) Design and performance analysis of a resorption cogeneration system. Int J Low-Carbon Technol 8:i85–i91
41. Bao H, Ma Z, Roskilly AP (2016) An optimised chemisorption cycle for power generation using low grade heat. Appl Energy 186:251–261
42. Wang L, Ziegler F, Roskilly AP, Wang R, Wang Y (2013) A resorption cycle for the cogeneration of electricity and refrigeration. Appl Energy 106:56–64
43. Lépinasse E, Marion M, Goetz V (2001) Cooling storage with a resorption process: application to a box temperature control. Appl Therm Eng 21:1251–1263
44. Vasiliev LL, Mishkinis DA, Antukh AA, Kulakov AG, Vasiliev LL (2004) Resorption heat pump. Appl Therm Eng 24:1893–1903
45. Jin ZQ, Wang LW, Jiang L, Wang RZ (2013) Experiment on the thermal conductivity and permeability of physical and chemical compound adsorbents for sorption process. Heat Mass Transf 49:1117–1124
46. Lu HB, Mazet N, Spinner B (1996) Modelling of gas-solid reaction—coupling of heat and mass transfer with chemical reaction. Chem Eng Sci 51:3829–3845
47. Wang R, Wang L, Wu J (2014) Adsorption refrigeration technology: theory and application. Wiley
48. Vijayaraghavan S, Goswami DY (2003) On evaluating efficiency of a combined power and cooling cycle. J Energy Res Technol 125:534–547
49. Bianchi M, Pascale AD (2011) Bottoming cycles for electric energy generation: parametric investigation of available and innovative solutions for the exploitation of low and medium temperature heat sources. Appl Energy 88:1500–1509

Index

© Science Press 2021
L. Wang et al., *Property and Energy Conversion Technology of Solid Composite Sorbents*, Engineering Materials,
https://doi.org/10.1007/978-981-33-6088-4

Printed in the United States
by Baker & Taylor Publisher Services